IA	IIA	IIIA	IVA	VA	VIA	VIIA	VIII			IB	IIB	IIIB	IVB	VB	VIB	VIIB	O
1 H 1.008																	2 He 4.003
3 Li 6.941	4 Be 9.012											5 B 10.81	6 C 12.01	7 N 14.01	8 O 16.00	9 F 19.00	10 Ne 20.18
11 Na 22.99	12 Mg 24.31											13 Al 26.98	14 Si 28.09	15 P 30.97	16 S 32.06	17 Cl 35.45	18 Ar 39.95
19 K 39.10	20 Ca 40.08	21 Sc 44.96	22 Ti 47.90	23 V 50.94	24 Cr 52.00	25 Mn 54.94	26 Fe 55.85	27 Co 58.93	28 Ni 58.71	29 Cu 63.55	30 Zn 65.38	31 Ga 69.72	32 Ge 72.59	33 As 74.92	34 Se 78.96	35 Br 79.90	36 Kr 83.80
37 Rb 85.47	38 Sr 87.62	39 Y 88.91	40 Zr 91.22	41 Nb 92.91	42 Mo 95.94	43 Tc 98.91	44 Ru 101.1	45 Rh 102.9	46 Pd 106.4	47 Ag 107.9	48 Cd 112.4	49 In 114.8	50 Sn 118.7	51 Sb 121.8	52 Te 127.6	53 I 126.9	54 Xe 131.3
55 Cs 132.9	56 Ba 137.3	57 La 138.9	72 Hf 178.5	73 Ta 180.9	74 W 183.9	75 Re 186.2	76 Os 190.2	77 Ir 192.2	78 Pt 195.1	79 Au 197.0	80 Hg 200.6	81 Tl 204.4	82 Pb 207.2	83 Bi 209.0	84 Po (210)	85 At (210)	86 Rn (222)
87 Fr (223)	88 Ra 226.0	89 Ac (227)															

Lanthanides														
57 La 138.9	58 Ce 140.1	59 Pr 140.9	60 Nd 144.2	61 Pm (147)	62 Sm 150.4	63 Eu 152.0	64 Gd 157.3	65 Tb 158.9	66 Dy 162.5	67 Ho 164.9	68 Er 167.3	69 Tm 168.9	70 Yb 173.0	71 Lu 175.4

Actinides														
89 Ac (227)	90 Th 232.0	91 Pa 231.0	92 U 238.0	93 Np 237.0	94 Pu (242)	95 Am (243)	96 Cm (247)	97 Bk (251)	98 Cf (254)	99 Es (253)	100 Fm (256)	101 Md (256)	102 No (254)	103 Lw (257)

Oxford Chemistry Series

General Editors

P. W. ATKINS   J. S. E. HOLKER   A. K. HOLLIDAY

Oxford Chemistry Series

1972
1. K. A. McLauchlan: *Magnetic resonance*
2. J. Robbins: *Ions in solution* (2): *an introduction to electrochemistry*
3. R. J. Puddephatt: *The periodic table of the elements*
4. R. A. Jackson: *Mechanism: an introduction to the study of organic reactions*

1973
5. D. Whittaker: *Stereochemistry and mechanism*
6. G. Hughes: *Radiation chemistry*
7. G. Pass: *Ions in solution* (3): *inorganic properties*
8. E. B. Smith: *Basic chemical thermodynamics*
9. C. A. Coulson: *The shape and structure of molecules*
10. J. Wormald: *Diffraction methods*
11. J. Shorter: *Correlation analysis in inorganic chemistry: an introduction to linear free-energy relationships*
12. E. S. Stern (ed): *The chemist in industry* (1): *fine chemicals for polymers*
13. A. Earnshaw and T. J. Harrington: *The chemistry of the transition elements*

A. EARNSHAW and T. J. HARRINGTON
UNIVERSITY OF LEEDS   BRADFORD GRAMMAR SCHOOL

The chemistry of the transition elements

Clarendon Press · Oxford · 1973

Oxford University Press, Ely House, London W.1

GLASGOW NEW YORK TORONTO MELBOURNE WELLINGTON
CAPE TOWN IBADAN NAIROBI DAR ES SALAAM LUSAKA ADDIS ABABA
DELHI BOMBAY CALCUTTA MADRAS KARACHI LAHORE DACCA
KUALA LUMPUR SINGAPORE HONG KONG TOKYO

PAPERBACK ISBN 0 19 855425 7
CASEBOUND ISBN 0 19 855435 4

© OXFORD UNIVERSITY PRESS 1973

PRINTED IN GREAT BRITAIN BY
J. W. ARROWSMITH LTD., BRISTOL, ENGLAND

Editor's foreword

VAST developments have occurred in transition-metal chemistry during the past two decades; new compounds, new applications and new theories of bonding have all contributed to the rapid growth of this area of chemistry. For the student, a balanced introduction is therefore all the more necessary, and this book is written to provide it.

Characteristics of transition metals generally (e.g. complex formation, variable oxidation state, magnetic properties) are described first; the relevant theories of bonding are then outlined, using a minimum of mathematics; finally, the individual elements are treated group by group, with an emphasis on trends, and with explanations based on the preceding sections.

From this treatment, the student should obtain an overall view of transition-metal chemistry, and so be able to proceed to more sophisticated or detailed treatments of particular topics, as provided by other books in this series. These include *Chelation*, *Lanthanides and actinides*, and *The ligand field*. Details of aqueous solution behaviour will be found in *Ions in solution* (3): *inorganic properties*, and for a broad introduction the student should read *The periodic table of the elements*.

A.K.H.

Preface

AN understanding of bonding theory is central to a study of the chemistry of any element, and this is particularly true of the transition elements, which exhibit a wider variety of bonding than is to be found in any other part of the Periodic Table. A considerable portion of this book is therefore devoted specifically to the various bonding theories, which are basic to the subsequent discussion of the chemistry of the individual elements. This chemistry can be systematically described by considering the elements either in 'vertical' groups (as is usual with main-group elements) or in 'horizontal' rows (as with the lanthanides and actinides). Each approach has its advantages and neither is entirely adequate on its own. We have chosen the former approach, but have attempted to retain some of the advantages of the latter by indicating the more important 'horizontal' similarities and trends in the preceding account of bonding theory.

It is a pleasure to record our thanks to the people who have helped in the preparation of this book—to Professor D. A. Long who suggested we write it; to Mr J. B. Bentley and Professor A. K. Holliday who read the manuscript and whose incisive comments prevented a number of errors; to the Sixth formers of Bradford Grammar School and the undergraduates of Leeds University on whom much of the material has been tried out; to the Universities of Birmingham, Bristol, Essex, Leeds, Oxford, Southampton, and Sussex for permission to quote some questions from recent examination papers; to the staff of the Clarendon Press for their helpful cooperation, and finally to our wives Jean and Helena for their forbearance throughout.

ALAN EARNSHAW
JOHN HARRINGTON

Acknowledgement
The authors wish to thank Dr R. J. Puddephatt for permission to reproduce Figs 0.1, 0.3, and 0.4, which appear in *The periodic table of the elements* (OCS 3).

Contents

Introduction

THE probability distribution of an electron in an atom can be described in terms of three quantum numbers: the principal quantum number n which can take any positive integral value; the azimuthal or orbital quantum number l which can take any positive integral value from 0 to $(n-1)$ inclusive; and the magnetic quantum number m which can take any integral value (positive or negative) from 0 to l inclusive. The region in space in which the electron is most likely to be found is called an *orbital*†, the shape of which depends on the value of l.

A further quantum number, the spin quantum number s, which can take values of $+\frac{1}{2}$ or $-\frac{1}{2}$, is necessary to identify a particular electron within an atom. It is a consequence of Pauli's exclusion principle (no two electrons in an atom can have the same set of four quantum numbers) that each orbital can accommodate a maximum of only two electrons, one of which will have $s = +\frac{1}{2}$ and the other $s = -\frac{1}{2}$. Such electrons are said to be *paired*, whereas electrons in different orbitals but with the same value of s are said to be *unpaired*.

When $l = 0$ (possible no matter what the value of n), the value of m must be 0 and a single, s, orbital results. This is spherically symmetrical about the nucleus as shown in Fig. 0.1(a).

When $l = 1$ (possible for $n \geqslant 2$), m can take any of three values, 0, -1, or $+1$, and the three resulting orbitals are known as p-orbitals. They are mutually perpendicular and are conventionally referred to as p_x, p_y, and p_z orbitals, each being composed of two lobes situated on the appropriate axis (Fig. 0.1(b)).

When $l = 2$ (possible for $n \geqslant 3$), m can take any of five values, -2, -1, 0, $+1$, or $+2$, giving rise to five d-orbitals (Fig. 0.1(c)). Three of these have four lobes each and lie between the axes. They are the d_{xy}, d_{xz}, and d_{yz} orbitals, collectively known as the t_{2g} or the d_ε set of orbitals. The fourth, known as the $d_{x^2-y^2}$ orbital, is similar to the first three except that its lobes lie along two axes, while the fifth, the d_{z^2} orbital, lies along the remaining axis. This last orbital can be regarded as a combination of $d_{y^2-z^2}$ and $d_{z^2-x^2}$ orbitals both analogous to the $d_{x^2-y^2}$. The $d_{x^2-y^2}$ and d_{z^2} orbitals are collectively known as the e_g or d_γ set of orbitals. Despite their different shapes all five d-orbitals have the same energy, i.e. are said to be *degenerate*, in an isolated atom.

When $l = 3$ (possible for $n \geqslant 4$), m can take any of the seven values -3 to $+3$ giving rise to seven f-orbitals. We need not be concerned further with these

---

† Strictly, an *orbital* is defined as the 'wave function' ψ of an electron, and the probability of an electron being in a particular place depends on $|\psi|^2$. The orbital diagrams given here are actually functions of $|\psi|^2$, not ψ, but the distinction is not important here.

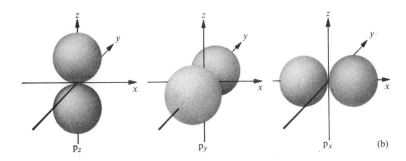

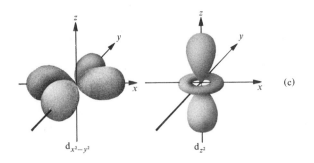

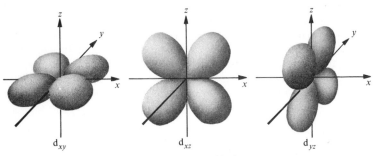

Fig. 0.1. Atomic orbitals.

except to note that, as with d-orbitals, a set of f-orbitals is degenerate in an isolated atom.

The various sets of atomic orbitals, however, differ in energy, and this variation depends primarily on which electron shell is under consideration ($n = 1, 2, 3$, etc.), but partly also on the type of orbital (s, p, d, or f) and partly on the number of electrons occupying the orbitals. The arrangement of electrons within the orbitals, the *electronic configuration*, is governed by Pauli's exclusion principle (already mentioned) and Hund's first rule (the most stable electronic configuration is that in which the maximum possible number of electrons remains unpaired).

The variation of the energies of atomic orbitals with increasing atomic number in neutral atoms is shown qualitatively in Fig. 0.2. The Periodic Table may be considered to be built up by the successive addition of electrons to the energetically lowest available orbital and a simultaneous addition of protons (and neutrons) to the nucleus. On this basis it can be seen that the orbitals are

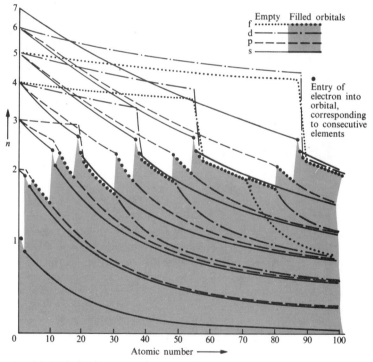

FIG. 0.2. Variation in energy of atomic orbitals with increasing atomic number.

filled in the order: 1s, 2s, 2p, 3s, 3p to produce the first and second short periods. The energies of these orbitals fall rapidly as the positive charge on the nucleus increases. Shielding of the nucleus by already filled orbitals is inefficient in these cases owing to the fact that s- and p-orbitals, even of high principal quantum number, penetrate the core of filled shells and extend close to the nucleus. This is not the case however for d-orbitals which, being less penetrating, are quite efficiently shielded from the nucleus by filled s- and p-orbitals of lower principal quantum number. As a consequence, when the 3p level has been filled (argon), the next lowest level is not 3d but 4s, which therefore takes the next two electrons to give potassium and calcium. The 3d orbitals are now partially enveloped by the filled 4s and their energy drops sharply. The subsequent filling of the 3d orbitals provides a series of ten elements before the 4p orbitals are filled to complete the first long period. The second long period is then produced in an analogous manner by the filling of the 5s, 4d, and 5p levels.

The next period however is complicated by the still empty 4f level. f-orbitals are even less penetrating than d-orbitals, and are thus more effectively shielded by core electrons. As a result, it is not until the 6s orbital has been filled (caesium and barium) that the energy of the 4f level is significantly reduced. Even so, the next electron enters the 5d level to give lanthanum before the seven 4f orbitals are filled, giving a series of fourteen *lanthanide* or 'rare-earth' elements. The filling of the 5d level is then completed, followed by the 6p. The same process is repeated in the final period as far as the heaviest atom yet discovered.

The periodic table is made up of three types of elements: s- and p-block, or *main-group* elements; d-block, or *transition* elements; and f-block or *inner transition* elements. Each period is completed by a noble gas, whose electronic configuration provides an inert core of electrons for the subsequent period. The chemistry of main-group elements is largely explained by their tendency to attain the configuration of the nearest noble gas by either losing or gaining electrons. This leads to the concept of a group valency† which equals the number designated to the group in the periodic table and differs by one between adjacent groups across a period.

With main group elements the greatest similarities occur vertically within each group. With inner-transition elements the most marked similarities occur horizontally right across the series, the chemistry of these elements, particularly the lanthanides, being dominated by their ready loss of the three $4d^1$ and $5s^2$ electrons to form M^{3+} ions. This ionic charge has the effect of lowering the energy of the 4f orbitals so much that they enter the inert core of electrons and do not participate further in chemical bonding.

† Valency is the ionic charge or the number of covalent bonds formed in chemical combination, and the outer-shell electrons participating are known as valence electrons.

The transition elements are intermediate between the main-group elements and the inner-transition types. At the beginning of each series the ns and $(n-1)d$ orbitals are of similar energy and scandium, yttrium, and lanthanum show a group valency of three, following naturally after the alkali and alkaline-earth groups. However, in proceeding across each series, the $(n-1)d$ orbitals fall in energy compared to the ns orbitals, and by the time zinc, cadmium, and mercury are reached they are firmly embedded in the inert electron core and are not involved at all in the chemistry of these elements. For this reason, these elements are frequently not considered as transition elements at all. However, for the body of transition elements, both ns and $(n-1)d$ electrons may be involved to varying extents in bond formation, leading to the 'transitional characteristics' discussed in the next chapter and producing both vertical and horizontal similarities. The relative energies of the orbitals involved are summarized in Fig. 0.3.

A further point to be mentioned at this stage is the *lanthanide contraction*. When a set of orbitals is gradually filled, the increased electrostatic attraction between the valence electrons and the necessarily increased nuclear charge causes the atom to contract. In the case of the lanthanide series, where an antepenultimate shell is being filled, it might be thought that the 4f electrons would shield one another and also the 6s and 5d valence electrons from the

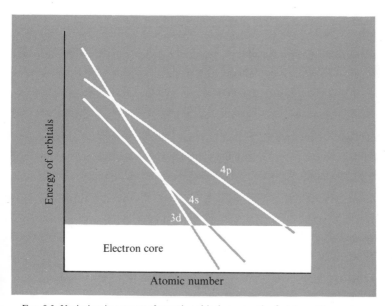

FIG. 0.3. Variation in energy of atomic orbitals across the first transition series.

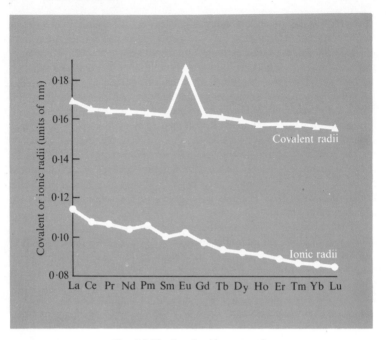

FIG. 0.4. The Lanthanide contraction.

increased nuclear charge, thus preventing any contraction. However, shielding by the 4f electrons is poor, and a small but (fairly) steady contraction occurs across the series. This is seen in both the atomic or covalent radii and the trivalent ionic radii (Fig. 0.4).

The lanthanide contraction has an important effect on the relationship between the three series of transition elements. There is a considerable increase in atomic and ionic radii in going from the first to the second series, but the contraction across the lanthanides almost exactly equals the expansion which would otherwise have occurred between the second and third transition series. Hence pairs of elements in the second and third series are often almost identical in size and have very similar chemical properties which differ markedly from the member of the family in the first series.

1. Some characteristics of transition elements

THE differences in electronic structure which distinguish the transition elements from all other classes have already been noted. These differences lead to physical and chemical properties which are characteristic of transition elements and may be considered under the following headings:

Variable oxidation state,
Complex formation,
Coloured compounds,
Magnetic properties,
Metallic nature,
Catalytic activity.

While it is not suggested that the individual properties to be discussed under these headings are necessarily unique to transition elements, collectively they add up to behaviour quite distinct from that shown by any other type of element.

Variable oxidation state

The *oxidation state* or *oxidation number* of a particular element in a compound is the formal charge ascribed to each atom of the element. Its value is arrived at by assuming that the oxidation state of free or uncombined elements is zero; that the oxidation state of oxygen (except in peroxides and super-oxides) is always -2, and that the oxidation state of hydrogen (except in metal hydrides) is always $+1$. In addition, certain elements almost always display the same oxidation state, e.g. $+1$ for alkali metals and -1 for halogens (except when combined with oxygen or another halogen). On this basis it can be seen that manganese in potassium permanganate, $KMnO_4$, is in the $+7$ oxidation state and is represented as $Mn(VII)$. The concept of oxidation state has the advantage over that of valency that it assumes nothing about the actual nature (ionic or covalent) or the multiplicity of the bonding.

As transition elements usually have several electrons in different orbitals of comparable but not equal energies (Fig. 0.3, p. 5) there is the possibility of a variable number of electrons participating in bonding and giving rise to a variety of oxidation states. The relative stabilities of the different oxidation states are determined by several factors, for example: the electronic structure (reflected in the ionization energies and ionic radii), the type of bonding (ionic or covalent, σ or π†), the stereochemistry, the lattice energy, the solvent, and the solvation energy. With so many factors involved, it is clear that comparisons between different elements must only be made with the greatest

† See pp. 39–40.

TABLE 1.1

Oxidation states of first transition-series elements

Sc	Ti	V	Cr	Mn	Fe	Co	Ni	Cu	Zn
								1	
		2	2	2	2	2	2	2	2
3	3	3	3	3	3	3	3		
	4	4		4					
		5							
			6						
				7					

of care. Table 1.1 shows those oxidation states of elements of the first series which have more than a transient existence in aqueous solution.

For a given transition element, oxidation states differ by 1 rather than 2 as in the main-group elements (where variable oxidation state depends generally upon the inert-pair effect†). The highest possible oxidation state is given by the total number N of ns and $(n-1)$d electrons. Table 1.2 shows the ionization energies of the transition elements, and the clear break that occurs after the loss of N electrons precludes any possibility of further electrons being involved in chemical bonding.

Subject to this limit on the number of electrons available, the formation of cations (as opposed to covalent oxidation states) is governed mainly by Fajans' rules. These state that between a cation and an anion, covalency

TABLE 1.2

Ionization energies of the transition elements of the first series (kJ mol^{-1})

	1st	2nd	3rd	4th	5th	6th	7th	8th
Sc	633	1234	2382	7120	8860	10 700	—	15 320
Ti	658	1308	2842	4170	9620	11 570	13 580	—
V	650	1412	2822	4630	6280	12 410	14 550	16 740
Cr	652	1398	2980	4820	7040			
Mn	717	1508	3248	5010	7320			
Fe	762	1560	2955					
Co	758	1645	3230					
Ni	736	1750	3382					
Cu	744	1955	3550					
Zn	905	1730	3825					

† This refers to the situation found in some p-block elements, where an oxidation number of 2 less than the group oxidation number sometimes occurs. This is because the ns^2 electrons, being more penetrating and so more firmly held by the nucleus than are the np electrons, are not in these cases involved in bond formation.

increases the higher the charge on either anion or cation, the smaller the cation, and the larger the anion. Consequently, as the oxidation number increases, so do the covalent characteristics, and we find oxides becoming more acidic and halides more easily hydrolysed†. In the first transition series, where the ionic radii are fairly small, this effect is particularly marked.

Except for copper, the lowest oxidation state which is not stabilized by π-bonding is $+2$. In the neutral atoms the energies of the ns and $(n-1)$d orbitals are similar, but once the atom becomes charged by losing one of the s-electrons, the energy of the d-orbitals is lowered and, as can be seen from the ionization energies (Table 1.2), the second s-electron is easily lost. In the case of copper, the stability of the symmetrical d^{10} configuration allows the loss of a single electron to form copper(I), though even here disproportionation into metallic copper and copper(II) readily occurs in aqueous solution unless a complexing agent is present to stabilize the copper(I).

In the first half of the series—up to manganese, technetium, and rhenium, which have the $(n-1)d^5 ns^2$ configuration—the highest oxidation state for any element is equal to the total of its $(n-1)$d and ns electrons. However, these oxidation states are attained with increasing difficulty and their stabilities decrease in progressing across each series. This is reflected in their increasing oxidizing power, so that for the first series Sc(III) < Ti(IV) < V(V) < Cr(VI) < Mn(VII), while Fe(VIII) is not known to occur at all. As we shall see, the higher oxidation states become more stable down each group, with the result that Ru(VIII) and Os(VIII) are known, but they are strongly oxidizing and this oxidation state is the highest displayed by any element. Subsequently the d-orbitals become more strongly entrenched in the inert electron core and by the time copper is reached $+3$ is the highest oxidation state attainable.

Along a series, the ionic radius changes only slightly for a given oxidation state, and therefore considerable 'horizontal' similarities occur, the $+2$ and $+3$ states providing the opportunities for the most extensive comparisons. However, the observed trends are not regular and this is largely a result of the extra stability of half-filled (d^5) and completely filled (d^{10}) d-orbitals. These configurations provide the most symmetrical distribution of electrons which suffer the minimum mutual repulsion and screening from the attraction of the nucleus. Oxidation states involving these arrangements of electrons consequently show an appreciably greater stability than would otherwise be

---

† The higher the oxidation state of the metal, the higher its net positive charge even though the bonds are increasingly covalent. In aqueous solution the charge may be neutralized by the formation of a covalent bond with the negatively charged hydroxyl ion, thus promoting the ionization of the water, $H_2O \rightleftharpoons H^+ + OH^-$. Oxides are thus acidic because of the H^+ produced and, in the case of the halides, replacement of halide ions by hydroxide or oxide (hydrolysis) is facilitated. For a particular oxidation state, the more electropositive the metal is, the more acceptable is the net positive charge and so the less acidic the oxide.

expected, though the effect is less pronounced in the second and third than in the first series. 4d and 5d orbitals are larger than 3d, so the mutual repulsion of electrons within them is smaller, and any reduction in repulsion proportionately less important.

In passing from the first to the second and then to the third series, the energy differences between the ns and the $(n-1)$d orbitals decreases, and the valence electrons are less strongly attracted to the nucleus as the size of the atom increases. This causes the higher oxidation states to become more stable, and the essentially ionic $+2$ and $+3$ states to become less important.

Fajans' rules suggest that the larger cations of the second and third series will give less-covalent compounds (i.e. less-acidic oxides) than those of the first series. This is undoubtedly true of the group oxidation states where the cations are without d-electrons and have essentially spherical, inert-gas structures. In lower oxidation states the presence of d-electrons complicates the simple considerations on which the rules are based and, where they occur, the lower oxidation states of the second- and third-series elements are in fact usually more covalent than their first-series analogues. Also, partly owing to the reduced interelectronic repulsions in 4d and 5d orbitals, spin pairing becomes more common with the heavier elements.

PROBLEM

1.1. Iron(III) ions oxidize iodide ions to iodine in weakly acid solution, but addition of iodine to a suspension of iron(II) hydroxide in alkali produces iron(III) hydroxide and iodide ions. Why?

Complex formation

Because of their small size and comparatively high nuclear or ionic charge, transition-metal cations exert strong electrostatic attractions on molecules or ions containing lone pairs of electrons (i.e. two non-bonding electrons in the same orbital). Such molecules or ions are known as *ligands* and the individual atom possessing the lone pair as the *donor atom*. The product is a *coordination compound* or *complex*. If it possesses an overall charge it is known as a complex anion or cation as the case may be. So widespread is this phenomenon of coordination that the solution chemistry of transition elements can only be understood on the basis of their coordination chemistry.

A ligand may be simply a monatomic anion, such as a halide ion, or it may be a polyatomic molecule or ion containing a donor atom from the oxygen or nitrogen groups. Indeed a ligand may contain more than one donor atom; a ligand with two donor atoms is said to be *bidentate*, and the prefixes tri, tetra, penta, and hexa replace bi to signify three, four, five, and six donor atoms respectively. For instance, ethylene diamine $H_2N \cdot CH_2 \cdot CH_2 \cdot NH_2$ is a bidentate ligand, and can coordinate to a single cation as shown in Fig. 1.1.

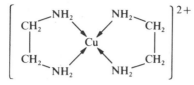

FIG. 1.1. The bis(ethylenediamine)copper(II) ion.

The term *coordination number* is used to denote the number of donor atoms associated with the central atom. This is clearly comparable to the use of the same term for the number of ions of opposite charge surrounding a particular ion in ionic crystals. Thus in the hexacyanoferrate(II) ion, $[Fe(CN)_6]^{4-}$, the coordination number of the iron atom is six and its oxidation number $+2$. The coordination number can vary from one element to another but, for complexes of transition elements in the $+2$ and $+3$ oxidation states, is usually four or six. It only occasionally exceeds six even for higher oxidation states, and this is confined largely to the first half of the series. Of course, the shape, or stereochemistry, of the complex is dependent on the coordination number and, from the point of view of the metal, is defined by the arrangement of donor atoms around it. Fig. 1.2 shows most of the stereochemistries which correspond to the different coordination numbers.

Where a particular complex contains a number of ions or molecules capable of acting as ligands, then isomerism is possible as different ligands enter the *coordination sphere*† (denoted by square brackets). A well-known example of this is found in the series of chromium(III) chloride hydrates $CrCl_3 \cdot 6H_2O$ obtained from aqueous solution (Table 1.3). The more free ions that are present, the greater is the electrical conductivity of solutions of the complex, and conductivity measurements therefore provide a very useful means of deciding whether chloride ions or water molecules are bound inside

TABLE 1.3

The chromium(III) chloride hexahydrates, $CrCl_3 \cdot 6H_2O$

Compound	Colour	Number of Cl^- ions	Total number of ions
$[Cr(H_2O)_6]^{3+}3Cl^-$	Violet	3	4
$[Cr(H_2O)_5Cl]^{2+}2Cl^-, H_2O$	Pale green	2	3
$[Cr(H_2O)_4Cl_2]^+Cl^-, 2H_2O$	Dark green	1	2

There is a fourth form, the ether-soluble, non-electrolyte $[Cr(H_2O)_3Cl_3]$ (obtained by Recoura) which instantly goes to the dark-green form in aqueous solution.

† There are many other sources of isomerism in complexes, and this interesting topic is discussed in detail in most general inorganic textbooks.

Coordination number	Stereochemistry		Example

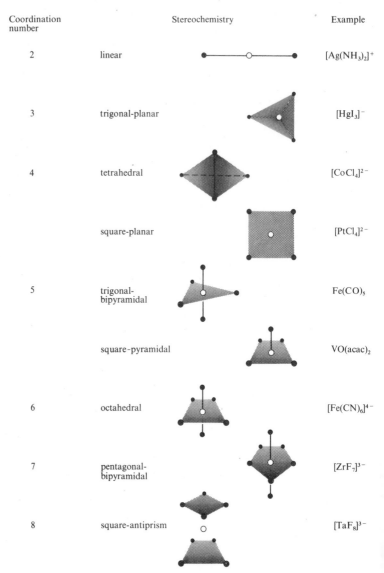

2	linear	$[Ag(NH_3)_2]^+$
3	trigonal-planar	$[HgI_3]^-$
4	tetrahedral	$[CoCl_4]^{2-}$
	square-planar	$[PtCl_4]^{2-}$
5	trigonal-bipyramidal	$Fe(CO)_5$
	square-pyramidal	$VO(acac)_2$
6	octahedral	$[Fe(CN)_6]^{4-}$
7	pentagonal-bipyramidal	$[ZrF_7]^{3-}$
8	square-antiprism	$[TaF_8]^{3-}$

FIG. 1.2. Coordination numbers and stereochemistries.

the coordination sphere. Also in this case, the number of chloride ions outside the coordination sphere can be determined, because they are the ones which are immediately precipitated by the addition of silver nitrate solution.

Various theories have been used to explain the coordinate bond and it is evident that several factors are involved—the simple electrostatic attraction between the metal ion and the ligand; σ-donation of electrons by the ligand to the cation, which will be facilitated by empty orbitals on the metal; π-donation to or from the metal (depending on the ligand, and often being made possible by the presence of partially filled metal d-orbitals). Each theory places a different emphasis on each of these factors and which theory is appropriate, and gives the most satisfactory results, will depend on the particular compound in question. However, when a transition metal loses its ns^2 electrons, the $(n-1)$d orbitals become partially exposed and will be influenced considerably by their environment. We need not be surprised therefore to see in the next chapter that complexing has a marked effect on the energies of these d-orbitals, causing the level to split, nor that otherwise unstable oxidation states (e.g. copper(I) or cobalt(III) in aqueous solutions) can be stabilized by suitable ligands.

PROBLEMS

1.2. $N(CH_2CH_2NH_2)_3$ is a tetradentate ligand. What stereochemistry does it favour and why?

1.3. A solution containing 0·32 g of the complex $CrCl_3 \cdot 6H_2O$ was passed through a cation-exchange resin in the acid form and the liberated acid was equivalent to 28·3 ml of 0·125 mol dm^{-3} sodium hydroxide solution. Deduce the structure of the complex. (*University of Sussex*)

1.4. Mercury(II) iodide dissolves freely in potassium iodide solution, and in the process the freezing point of the solution is raised. Explain.

Coloured compounds

Colours and variations in the colours of compounds can be produced simply by the physical effect of having the particle size in the range of the wavelengths of visible light (approximately 400–700 nm), but for the chemist a much more important source of colour is the absorption of visible light. For absorption to occur, a substance must be capable of an excitation involving an amount of energy E which can be provided by the light. This can only be done by light of a particular frequency v, defined by $E = hv$ where h is Planck's constant. For this frequency to correspond with that of visible light, only the energies of electronic transitions are of the correct order of magnitude, though these also extend beyond the visible into the infrared and ultraviolet. In particular, the splitting of the d-orbitals brought about by complexing generally involves an energy difference of the magnitude required for the absorption of visible light.

In this context, it is worth noting that the 'coloured monatomic cations' characteristic of the transition elements are almost invariably complex ions. Typical is the hexa-aquotitanium(III) ion $[Ti(H_2O)_6]^{3+}$ responsible for the reddish-violet colour of aqueous solutions of titanium(III) salts. The coordination of the six water molecules to the Ti^{3+} ion splits its d-orbitals, as we shall see later (p. 26), into t_{2g} and e_g (or d_ε and d_γ) levels. The single unpaired electron of the titanium then absorbs green light of wavelength about 500 nm as it 'jumps' or is promoted from the t_{2g} to the e_g level†, producing the complementary colour reddish-violet. After this absorption, the single electron in the e_g level is of course unstable and will subsequently return to the more stable t_{2g} level with the loss of energy. This energy is released not as light but as translational energy (i.e. heat) among the neighbouring ions. The comparatively feeble colours of the hexa-aquomanganese(II) ion (pink) and the hexa-aquoiron(III) ion (amethyst) are readily explained. Both these are d^5 ions, and promotion of an electron within the d-orbitals inevitably results in the pairing of electron spins, which diminishes the probability of the transition.

There is another source of colour which must be mentioned briefly. Many transition-metal compounds show very intense colours, and the strong absorption bands vary with the oxidizing or reducing nature of the ligands. With a given metal cation, the bands move to longer wavelength as the ligand becomes easier to oxidize or reduce, and the colour in this case is due not to d-orbital transitions of the type mentioned above, but to the more-or-less complete transfer of an electron from the metal atom to the ligand or vice versa. The energy required for the transfer is obtained by the absorption of light of the appropriate frequency, and in the return transfer the energy is released as before in the form of heat. Typical examples are the ions MnO_4^-, TcO_4^-, and ReO_4^-. The metal is formally in the $+7$ oxidation state and is a ready acceptor of electrons. The higher oxidation state becomes more stable with increasing atomic number, and hence the permanganate (where the M^{VII} is the most oxidizing) absorbs at the lowest frequency, since the electron transfer takes place most easily in its case. The absorption moves towards the violet for the other ions, and hence the observed colours are purple (MnO_4^-), dark red (TcO_4^-) and colourless (ReO_4^-). The intense colour of the charge-transfer spectra can also be observed when the metal is in a low oxidation state and can transfer electrons to the ligand. The phenanthroline complexes of copper(I) and iron(II) are typical, and many analytical colour tests such as the formation of the well-known blood-red iron(III)-thiocyanate complex depend on the intense colour produced by charge transfer. The transfer of

† This treatment of the origin of spectra on the basis of a simple one-electron transition is satisfactory for d^1 and also for d^9 ions (the latter configuration is only one electron short of a completely filled d-level and absorption may be treated as the promotion of a 'positive hole'). However, it is only fair to point out that the position is much more complicated for all other configurations.

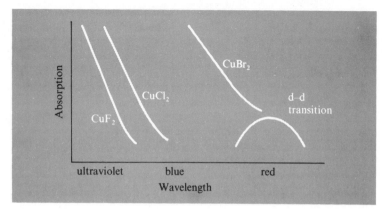

Fig. 1.3. Absorption spectra (idealized) of copper (II) halides. The total spectrum of each compound is a combination of the charge-transfer and d–d bands.

electrons might in certain cases take place very easily indeed, and result in a permanent redox reaction occurring. The deep colours of anhydrous copper(II) and iron(III) chlorides and bromides show that the charge-transfer bands are at long wavelength, and it is not surprising to find that the corresponding iodides have at best only transitory existence, breaking down to free iodine and the iodides of lower oxidation states of the metals (CuI and FeI$_2$). The absorption bands for the copper(II) halides are summarized in Fig. 1.3.

Magnetic properties

Electrons possess charge and, in atoms, are in motion with respect to the nucleus. They therefore behave as electric currents and set up associated magnetic fields. Electronic motion is more complicated than is suggested by the simple solar model of the atom, in which electrons are thought to orbit about the central nucleus and simultaneously to spin about their own axis. Nevertheless, even on the basis of more advanced quantum theory, an electron gives rise to two magnetic effects; one associated with its orbital quantum number l, and the other with its spin quantum number s. Together, these effects confer on each electron the properties of a small bar magnet with a magnetic moment. This moment interacts with any externally applied magnetic field. In filled electron shells the magnetic moments of the electrons neutralize one another, and the atom as a whole will only possess a resultant magnetic moment if unpaired electrons are present in the valence shell. Where this occurs, application of an external magnetic field will tend to cause the moments of the atoms to align themselves in the direction of the field. Such behaviour is known as *paramagnetism*, and paramagnetic substances are attracted into regions of strong magnetic field, the magnitude of the attraction being dependent on the number of unpaired electrons in the atom. A paramagnetic atom or molecule is characterized by its *effective magnetic moment*, μ_{eff}, which is usually expressed numerically as a 'Bohr magneton number'.†

In fact all substances are affected by an applied magnetic field whether they contain unpaired electrons or not. This is because the field will distort the motion of all electrons, effectively inducing small electric currents. Lenz's law of electricity states that currents induced magnetically are always such as to set up magnetic fields opposing the field producing them. This causes the substance to be repelled away from the regions of strong field, and the behaviour is known as *diamagnetism*. It is a weaker effect than paramagnetism and when unpaired electrons are present the paramagnetic effect swamps the diamagnetism.

It is evident from Hund's rule that the possession by an atom of degenerate and only partially filled d-orbitals will result in the presence of unpaired electrons. For this reason, paramagnetism is common among the compounds of transition and inner-transition elements whereas the compounds of main-group elements are almost invariably diamagnetic (O_2 and NO are notable exceptions).

A third type of magnetic behaviour is ferromagnetism, which is really very rare but of immense importance. It is a special case of paramagnetism and occurs in substances which contain a very high proportion of atoms or ions with unpaired electrons. In favourable circumstances the unpaired electrons of each atom interact and align themselves with the unpaired electrons of neighbouring atoms, the process being repeated throughout the material. The effect is thus on a macro rather than an atomic scale, and allows the construction of permanent magnets. Ferromagnetic behaviour is found mainly among the metals, alloys, and oxides of transition, and to a lesser extent inner-transition, elements.

A chemist is concerned in understanding the structure and bonding of compounds and it is for the compounds of the transition elements that these are most sensitively connected with magnetic properties. It is here that changes in structure and bonding are expected to produce the most obvious changes in magnetic behaviour. Main-group elements, as has been mentioned, produce compounds which are almost invariably diamagnetic. Lanthanides, though paramagnetic, show little variation in oxidation state and their unpaired electrons are quite well shielded by s- and p-electrons from the effects of other atoms in their compounds. In contrast, transition elements exhibit a wide

---

† One Bohr magneton (B.m.) is the magnetic moment thought by Bohr to be associated with the single electron in a hydrogen atom. Other magnetic moments were then expected to be simple multiples of this. This turned out to be an over-simplification, but the unit is still retained.

variety of oxidation states and their unpaired electrons are in d-orbitals exposed to their immediate surroundings. Changes in oxidation state, and in the magnitude of the d-orbital splitting produced by complexing, both affect the number of unpaired electrons and are paralleled by changes in the magnetic behaviour.

In complexes of the first transition series, μ_{eff} arises mainly from the magnetic effect associated with the *spin* of the electrons, the movement of the electrons which produces the *orbital* contribution being largely suppressed or *quenched* by the electrostatic field of the ligands. In these cases the Bohr magneton number is given by the *spin-only* formula:

$$\mu_{\text{eff}} = \sqrt{\{n(n+2)\}},$$

where n is the number of unpaired electrons. For elements of the second and third series the spin and orbital effects interact with each other in a rather complicated way and no such simple expression is applicable. The connection between n and the structure of a complex will be seen more fully when the theories of bonding are dealt with in the next chapter.

Metallic properties

Perhaps the most striking feature of the transition elements is that they are all metals, with high melting and boiling points, and good thermal and electrical conductivities. They are generally hard and strong, and form alloys with one another. It is the possession of these properties which gives the transition elements their unique technological importance.

The valence electrons of the transition elements are held rather loosely by the atomic nuclei, as is shown by their low ionization energies, and the atoms have no great affinity for further electrons. Bonding between pairs of metal atoms will thus be weak and we would not expect the pure elements to exist as, say, diatomic molecules. Instead, stability is achieved by the sharing of the valence electrons between many atoms so that each electron benefits from the attraction of several positive nuclei. It is a characteristic feature of the structure of a metallic solid that it exists as a lattice arrangement of positive ions held together by *delocalized* electrons permeating the whole lattice. For the transition metals, these electrons are provided from the atomic ns and $(n-1)d$ orbitals and the lattice is so formed that each ion usually has either eight or twelve similar nearest-neighbour ions. The electrons are extremely mobile and allow high electrical conductivity if a potential difference is applied across the metal.

The melting points (and boiling points) of the elements in all three transition series rise to a maximum at about group VI (Cr, Mo, and W), as shown in Fig. 1.4. The highest densities occur in almost the same region—osmium is the densest element. These facts show that the highest binding energies are achieved in the middle of the series, suggesting that these elements are

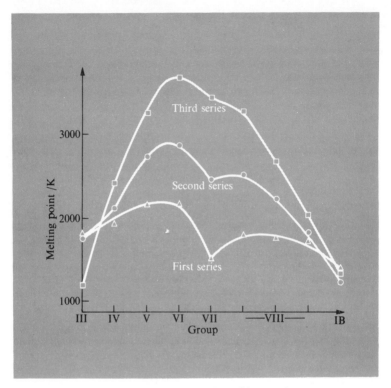

Fɪɢ. 1.4. Melting points of transition metals.

prepared to release the largest number of valence electrons for metallic bonding. It is notable that the same elements readily release electrons to oxidizing agents and so exhibit the highest oxidation states found in the periodic table. This correlation between binding energy and available oxidation states is common in metal chemistry.

The presence of delocalized electrons and the fact that all the ions are positively charged conveys a further property on the metallic lattice which distinguishes it from the lattice of an ionic salt. This is the ability of layers of ions to slide easily over one another on a 'cushion' of electrons without the necessity of overcoming strong electrostatic forces. Deformation by hammering and rolling are therefore comparatively easy.

Replacement of the metal atoms by others of similar size causes little distortion of the lattice and allows the production of alloys, which are usually solid solutions of variable composition and differing physical properties. Alternatively, *interstitial* compounds are formed by the incorporation, again in variable proportion, of small non-metallic atoms in the spaces between the metal atoms—the *interstices* of the lattice. Provided the incoming atoms are sufficiently small they do not completely disrupt the lattice but distort it somewhat and make it more rigid. The products are frequently notable for their extreme hardness and high melting point. In this respect they are often similar to ceramic materials, though their electrical and thermal conductivities are more metallic in character. The nature of the bonding is only partly understood and is still the subject of controversy. It is not fruitful to ascribe formal oxidation states to the metals in these materials. Examples are the hydrides, carbides, borides, and nitrides of transition metals, of which the carbides are of immense importance in the manufacture of steels of all types.

Catalytic activity

Nearly all transition elements have catalytic power, either as free elements or as compounds. The catalytic power of the metals themselves probably arises either from the use of their d-orbitals or from the formation of interstitial compounds to absorb and activate the reacting substances. Catalysis by metal compounds results from their ability to provide low-energy pathways for reactions either by facile change of oxidation state or by the formation of appropriate intermediates.

Catalysis is a complicated and incompletely understood phenomenon and generalization is dangerous. Suffice it to say that catalysts derived from transition elements are vital in many biological systems and indispensable in the chemical industry.

PROBLEM

1.5. The oxidation of iodide ions to iodine by peroxodisulphate(VI) ions is catalysed by both iron(II) and copper(II) ions. Suggest a possible reason for the catalytic action of these two ions. Which ion would you expect to be the more effective catalyst? Which other metallic ions might catalyse the reaction?

2. Theories of bonding

FOR most of the main-group elements of the Periodic Table, bonding is accomplished by the transfer or sharing of electrons in such a way that in attaining the group oxidation state the atoms also attain the stable electronic structure of the nearest noble gas. For this to occur in the case of the transition elements would require some impossibly high oxidation states. Even in the first half of the series we have seen that the group oxidation state becomes progressively less stable as it increases, and in any case many other oxidation states occur. This difficulty may be resolved for complexes if it is assumed that the coordinate bond is formed by the sharing of the lone-pair electrons of the ligand by both ligand and metal. Given the correct number of ligands, the metal thereby attains a noble-gas configuration without changing its oxidation state. Though this is a rather naïve view of the coordinate bond and is no longer accorded the significance it once was, it is still of value in understanding the carbonyls and organometallics.

The main theories which have been used to explain the coordinate bond are:

Valence-Bond (VB) theory, which assumes the bond to be covalent;

Crystal-Field (CF) theory, which assumes the bond to be purely electrostatic;

Ligand-Field (LF) theory, which is basically CF theory, but with some allowance for a covalent contribution; and

Molecular-Orbital (MO) theory, which allows for varying ionic and covalent contributions, but has the disadvantage of being difficult to apply rigorously. These theories will now be considered in turn.

Valance-bond theory

This approach, developed largely by Linus Pauling, was the one most widely used by chemists up to the mid nineteen fifties. Though now rarely used in coordination chemistry, it undoubtedly stimulated much of the work responsible for our present understanding of this field.

A coordinate bond is assumed to be formed by the donation of a lone pair of electrons from the donor atom of the ligand to an appropriate empty orbital on the metal. This can be illustrated by the simple case of the tetra-amminezinc(II) ion $[Zn(NH_3)_4]^{2+}$. The electronic configuration of the zinc atom is $4s^2\, 3d^{10}$, and of the zinc ion $3d^{10}\, 4s^0$. The 4s and 4p orbitals of the Zn^{2+} are therefore the orbitals of lowest energy which are empty, and so able to accommodate the four lone pairs from the ligands. The four metal–ligand bonds are equivalent, and this equivalence is achieved by the *hybridization*

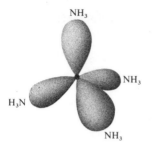

(a)

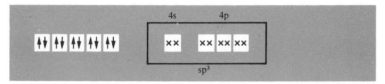

sp³

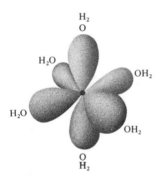

(b)

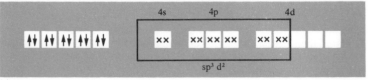

sp³ d²

↑↓ Represent non-bonding metal electrons   ×× Represent bonding lone-pair electrons

Fig. 2.1. (a) $[Zn(NH_3)_4]^{2+}$ (b) $[Zn(H_2O)_6]^{2+}$.

of the 4s and 4p orbitals to produce four sp^3 hybrid orbitals.† These point to the corners of a tetrahedron and form σ-(sigma) bonds‡ by overlapping with the lone-pair orbitals of the ligands, giving the complex ion a tetrahedral shape (Fig. 2.1(a)).

The hexa-aquozinc(II) ion is more complicated. If the Zn^{2+} ion is to accommodate the lone pairs from six water molecules it is necessary to involve two 4d orbitals in the bonding scheme. The resultant sp^3d^2 hybridization gives an octahedral shape to the ion (Fig. 2.1(b)).

The stepwise process of forming a complex ion may thus be considered hypothetically as:

  (i) the loss of electrons by the metal in accordance with its oxidation number,
 (ii) the hybridization of the appropriate metal orbitals which will define the stereochemistry of the complex,
(iii) the occupation of the hybrid orbitals by the lone pairs of electrons from the ligands to form σ-bonds.

In the case of zinc, the 3d electrons present no problem, but, as will be seen later, occasions arise when the requirements of bonding conflict with the tendency of non-bonding electrons to occupy all the 3d orbitals in accordance with Hund's rule. It is fundamental to the VB approach that in such cases the bonding electrons take precedence and the non-bonding electrons arrange themselves in whatever orbitals remain after bonding has occurred. Table 2.1

TABLE 2.1

Some common hybridizations and their geometries

Hybridization	Geometry
sp	linear
sp^2	planar
sp^3	tetrahedral
sd^3	tetrahedral
dsp^2 or sp^2d	square planar
dsp^3	trigonal bipyramidal (with d_{z^2}) or square pyramidal (with $d_{x^2-y^2}$)
d^2sp^3 or sp^3d^2	octahedral

† Hybridization is the mathematical combination of atomic orbitals to produce the same number of new orbitals each of the same shape and energy. The spatial distribution of the hybrid orbitals is determined by the particular atomic orbitals involved. For further details see C. A. Coulson: *The shape and structure of molecules* (OCS 9).

‡ A σ-bond is one in which the maximum electron density occurs on the axis between the two bonded atoms. The orbitals overlap 'end-on'.

shows the stereochemistries associated with various types of hybridization. In principle it should be possible to predict the stereochemistry of a complex on the basis of the available empty metal orbitals. However, this can lead to quite erroneous conclusions, and it is undoubtedly an advantage if the stereochemistry is known initially.

The possibility of using 4d orbitals for bonding, as for the $[Zn(H_2O)_6]^{2+}$ above, introduces complications in six-coordinate complexes of metals which have fewer 3d electrons. In such cases a choice of hybrid orbitals is available. If the 4d orbitals are used, the non-bonding electrons are arranged in the 3d orbitals in the same way as in the uncomplexed ion. If, on the other hand, 3d orbitals are used for bonding, then the non-bonding electrons may be forced to 'bunch' together in the remaining 3d orbitals. Complexes of the former type are known as 'outer-orbital' complexes, and those of the latter as 'inner-orbital' complexes. The hybrid orbitals are represented as sp^3d^2 and d^2sp^3 respectively. An obvious distinction is in the number (n) of unpaired electrons which result in the two cases. Clearly the use of 3d orbitals for bonding will reduce this number (unless there are fewer than three 3d electrons) and lead to a lower magnetic moment (Pauling's magnetic criterion of bond type). The situation is well illustrated by iron in its oxidation states of $+2$ and $+3$ (Fig. 2.2).

In the case of four-coordinate complexes, the use of different orbitals for bonding leads to either a tetrahedral (sp^3) or a square-planar (dsp^2) structure. The problem is particularly acute in the case of copper(II) complexes such as $[Cu(NH_3)_4]^{2+}$, since the number of unpaired electrons is the same in both cases, (Fig. 2.3). By analogy with zinc(II), sp^3 hybridization giving a tetrahedral

	3d	4s	4p	4d	n
Fe	↑↓ ↑ ↑ ↑ ↑	↑↓	□ □ □	□ □ □ □ □	4
Fe^{2+}	↑↓ ↑ ↑ ↑ ↑	□	□ □ □	□ □ □ □ □	4
sp^3d^2	↑↓ ↑ ↑ ↑ ↑	x x	x x x x x x	x x x x	4
d^2sp^3	↑↓ ↑↓ ↑↓ x x x x	x x	x x x x x x		0
Fe^{3+}	↑ ↑ ↑ ↑ ↑	□	□ □ □	□ □ □ □ □	5
sp^3d^2	↑ ↑ ↑ ↑ ↑	x x	x x x x x x	x x x x	5
d^2sp^3	↑↓ ↑↓ ↑ x x x x	x x	x x x x x x		1

FIG. 2.2. Electronic configurations of six-coordinate complexes of iron.

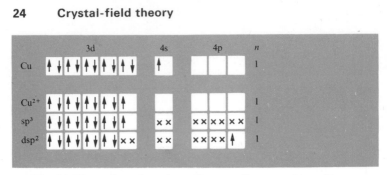

FIG. 2.3. Electronic configurations of four-coordinate complexes of copper.

structure might be expected. In fact, this is incorrect in the case of the tetram-mine, which is square-planar as indeed are most copper(II) complexes. A further objection to the interpretation given in Fig. 2.3 is that an odd electron in an 'exposed' orbital, such as the 4p is in this case, would be expected to be readily lost giving a copper(III) complex. But this oxidation does not easily occur. It has been suggested that the unpaired electron remains in the 3d level, and that a 4d orbital is used for bonding (sp^2d), but this seems to be a rather contrived way of avoiding sp^3 hybridization.

The disadvantages of VB theory can be summarized as follows.
  (i) It describes what has occurred, but does not explain why.
 (ii) In the case of d^7, d^8, and d^9 configurations (when inner d-orbitals are used for bonding) it requires promotion of electrons to largely unspecified orbitals.
(iii) While explaining to a certain extent the magnetic properties of complexes, it gives no explanation at all of their spectral properties. Yet, as both are dependent on the valence electrons of the metal, the two sets of properties must be related.

In a large number of cases, the crystal-field theory goes a considerable way to remedying these defects.

Crystal-field theory

This theory was developed at about the same time as the VB theory and was applied by physicists to transition-metal ions in a crystal lattice. The effects on the metal of the surrounding ions were assumed to be entirely electrostatic, and were shown to consist primarily of the splitting of the previously degenerate d-orbitals. The major part of this effect arises from the nearest-neighbour ions, even in an extended lattice. There is thus every reason to expect similar behaviour from a metal ion surrounded by negatively charged ligands. Even if the ligands are neutral, the lone pairs of electrons on the donor atoms will provide a localized negative charge which will interact with the metal. Nevertheless, the complete neglect of any covalent contribution was

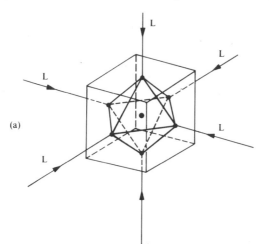

(a)

(b)

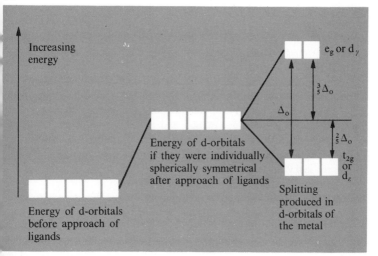

Fig. 2.4. (a) The hypothetical formation of an octahedral complex. (b) The resultant splitting of the metal d-orbitals.

difficult for chemists to accept, and it is certainly the major weakness of the theory. For this reason, and no doubt also because of the success initially enjoyed by the VB theory, this approach was not seriously applied by chemists until the nineteen fifties. By then it had become desirable to look for some explanation of the spectra of transition-metal complexes.

To see how the splitting of the d-orbitals arises, let us consider the hypothetical formation of a six-coordinate complex. These are invariably octahedral, and are the most widely investigated class of complex. The ligands are assumed to approach the metal ion along the three mutually perpendicular x-, y-, and z-axes. It is sometimes helpful to imagine the octahedron inscribed inside a cube, as in Fig. 2.4(a). The ligands, whether anionic or neutral, are considered to carry a negative charge by virtue of their lone-pair electrons which are directed towards the metal. As the ligands approach the metal ion, the d-electrons are repelled by the negative charge and the energy of the d-orbitals increases.† If each d-orbital were spherically symmetrical, they would all be repelled equally and would remain degenerate as in the isolated atom. But they are not individually spherically symmetrical. The axes of the e_g orbitals coincide with the x-, y-, and z-axes along which the ligands are approaching. They therefore suffer a greater repulsion and hence a greater increase in energy than do the t_{2g} orbitals which point between the axes (towards the centres of the edges of the circumscribed cube). When the ligands have reached their final positions, the separation in energy between the levels is defined as Δ_o. The e_g level is now higher in energy than would be the case if the metal ion were spherical, but the t_{2g} is lower. Their average, bearing in mind the different numbers of orbitals comprising the two levels, will remain the same as it would be if the metal orbitals were individually spherically symmetrical, but will of course be higher than it was in the metal ion before the ligands approached. This is shown diagrammatically in Fig. 2.4(b).

The importance of this splitting of the d-orbitals is in its effect on the electrons occupying them. These are now subject to two constraints:

(i) the tendency to remain unpaired and to occupy as many orbitals as possible, thereby reducing or eliminating the mutual repulsions‡ that occur when electrons are forced to pair in the same orbital,

(ii) the tendency of electrons to occupy the orbitals of lowest energy, in this case the t_{2g}.

The results of these constraints are illustrated in Fig. 2.5.

The high-spin and low-spin arrangements correspond to the outer- and inner-orbital descriptions of VB theory. In fact, the inner bonding orbitals of VB theory are the same e_g orbitals avoided by non-bonding electrons in CF theory. A discrepancy is seen however in the d^7 case. Whereas the inner-orbital configuration of VB theory uses the e_g orbitals for bonding and

† In discussing the destabilizing effects of the ligands' lone pairs, one should not overlook the fact that it is mainly the attraction between these electrons and the nuclear charge of the metal which gives the complex its stability.

‡ Mutual repulsions are easily understood, but are only part of the story. The *exchange* energy associated with electrons which are able to occupy several orbitals is actually the main reason for the applicability of Hund's rule, but this is less easily visualized.

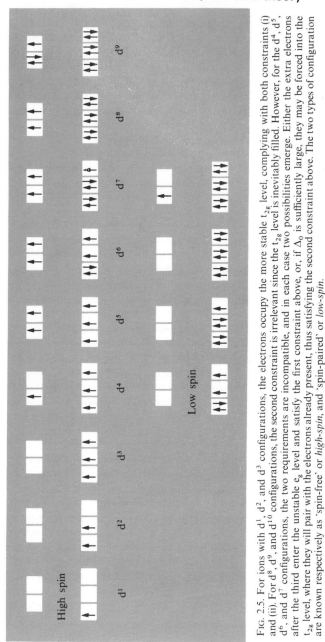

FIG. 2.5. For ions with d^1, d^2, and d^3 configurations, the electrons occupy the more stable t_{2g} level, complying with both constraints (i) and (ii). For d^8, d^9, and d^{10} configurations, the second constraint is irrelevant since the t_{2g} level is inevitably filled. However, for the d^4, d^5, d^6, and d^7 configurations, the two requirements are incompatible, and in each case two possibilities emerge. Either the extra electrons after the third enter the unstable e_g level and satisfy the first constraint above, or, if Δ_0 is sufficiently large, they may be forced into the t_{2g} level, where they will pair with the electrons already present, thus satisfying the second constraint above. The two types of configuration are known respectively as 'spin-free' or *high-spin*, and 'spin-paired' or *low-spin*.

consequently requires the promotion of a non-bonding electron to some higher orbital, no such promotion is needed by CF theory, and the electron occupies one of the e_g orbitals.

The use of magnetic measurements to estimate the number of unpaired electrons and so distinguish between the possible oxidation states and types of octahedral complex is still valid, and in addition, as we saw in Chapter 1, the colours of the complexes can be explained on the basis of electronic transitions within the d-level. The electronic spectra of d^1 and d^9 ions consist of a single absorption due to the promotion of one electron from the t_{2g} to the e_g level. To understand the spectra of other transition-metal ions requires a more detailed consideration of the energy levels arising from the interaction of the spin and orbital motion of electrons. Such a treatment is not possible within the simple framework of orbital splitting presented here, though it is still crystal-field theory, being based on the assumption of purely electrostatic bonding.

The formation of four-coordinate, tetrahedral complexes can be imagined to occur in much the same way as octahedral complexes. If the tetrahedron is inscribed inside a cube whose edges are parallel to the x-, y-, and z-axes, the ligands will approach towards four of the corners of the cube as shown in Fig. 2.6(a). If the d-orbitals are superimposed on this, it will be seen that the t_2† orbitals point towards the centre of the cube's edges, while the e point to the centre of the cube's faces. The t_2 electrons will now suffer greater repulsion by the ligand lone pairs than will the e electrons. The splitting of the d-orbitals will therefore be the reverse of that in the octahedral case, and is defined as Δ_t. This is shown in Fig. 2.6(b), and again is related to the 'average' energy of the d-orbitals. Δ_t will obviously be less than Δ_o in comparable cases, not only because four rather than six ligands are involved, but also because in the tetrahedral case none of the repulsions occur 'head-on' along the axes of the orbitals. In fact, for the same metal ion and ligands, it can be calculated that $\Delta_t = \frac{4}{9}\Delta_o$.

The magnitude of Δ (Δ_o or Δ_t) can be estimated by examination of the spectrum of the complex and will depend upon the ligand and metal involved, and also on the oxidation state of the metal.

For a particular metal in a given oxidation state, an increase in Δ means that more energy is required to bring about electronic promotion from one level to another. This causes a shift in the absorption in the direction red $\longrightarrow$ violet, while the observed colour of the complex changes in the opposite direction. By observing the spectra of complexes of the metal with a variety of ligands, it is possible to arrange the latter according to the magnitude of the splitting they produce. It is found that, with few exceptions, this order is the

---

† For reasons of terminology which need not concern us here, the subscript g is dropped for tetrahedral symmetry.

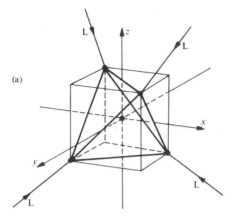

(a)

(b)

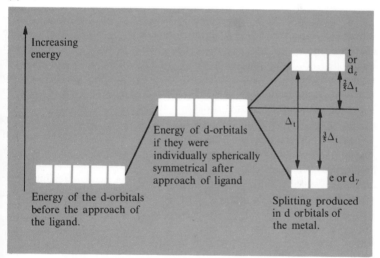

FIG. 2.6. (a) The hypothetical formation of a tetrahedral complex. (b) The resultant splitting of the metal d-orbitals.

same for most transition-metal ions, at least in the oxidation states $+2$ and $+3$. The list of ligands placed in order of increasing Δ is known as the Fajans–Tsuchida *spectrochemical series*:

$$I^- < Br^- < (\leftarrow SCN)^- \sim Cl^- < NO_3^- < F^- < urea \sim OH^- \sim$$
$$(\leftarrow ONO)^- \sim HCO_2^- < C_2O_4^{2-} < H_2O < (\leftarrow NCS)^- < EDTA^{4-} <$$
$$py \sim NH_3 < en < bipy < phen < (\leftarrow NO_2)^- \ll CN^- \sim NO^+ \sim CO$$

The abbreviations above are commonly used for the ligands they represent:

EDTA = ethylenediaminetetra-acetate

$$(^-O_2C\cdot CH_2)_2N\cdot CH_2\cdot CH_2\cdot N(CH_2\cdot CO_2{}^-)_2$$

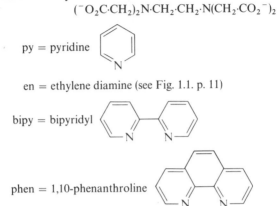

py = pyridine

en = ethylene diamine (see Fig. 1.1. p. 11)

bipy = bipyridyl

phen = 1,10-phenanthroline

The order above can be explained to some extent on the basis of other characteristics of the ligands,† but the position of ligands such as carbon monoxide is quite inexplicable unless, contrary to the basic assumption of crystal-field theory, an appreciable degree of covalency is admitted. This would be ligand-field theory.

Increasing the formal charge (oxidation state) on the metal will also increase Δ. The higher this charge the closer the ligand is attracted to the metal, and so the greater the repulsion experienced by the d-orbitals of the metal. $[Co(NH_3)_6]^{2+}$ is a high-spin paramagnetic complex, whereas $[Co(NH_3)_6]^{3+}$ is low-spin and diamagnetic.

The effect of different metals on the magnitude of Δ is not so readily seen. Along a given series the metal itself is of secondary importance compared to the ligand and the oxidation state. However, within a group Δ increases by about a third in moving from the first series to the second, and again from the second to the third. This to a large extent is due to the higher nuclear charges and the correspondingly greater attractions they exert on the ligands.

A major achievement of crystal-field theory is its explanation of the variation across a transition series of several thermodynamic properties. One of the properties for which the most complete sets of data are available is the hydration energy of dipositive ions of the first series. Hydration energy is defined

† For example, the acid ionization constants for the halogen hydrides in aqueous solution are in the order HI > HBr > HCl > HF. It seems reasonable to suppose that the greater the ability of X^- to associate with H^+, the greater the ability to associate (i.e. coordinate) with metal cations, and hence the greater Δ. However, carbon monoxide has no measurable affinity for protons and on this basis would be expected to be a very poor ligand.

as the heat evolved in the reaction:

$$M^{2+}(g) + \infty H_2O \rightarrow M^{2+}(aq).$$

The interaction of water molecules with different cations of the same charge will depend critically on the ionic radii of the latter. If the d-electrons of the transition metals were arranged spherically around the nucleus, a steady decrease in ionic radius and a steady increase in hydration energy would be expected across the series. Instead, a curve with two maxima is actually observed (Fig. 2.7). This can be understood on the basis of the Crystal-Field Stabilization Energies (CFSE) of the divalent ions. This is the stabilization accruing to the ions by virtue of the fact that the d-orbitals are split in an electrostatic field because they are not individually spherically symmetrical.

We now see the importance of the 'average' energy of the d-orbitals shown in Figs. 2.4(b) and 2.6(b). In an octahedral complex, a d^1 ion increases its stability by $\frac{2}{5}\Delta_o$ as compared to a spherical ion. For a d^2 ion the increase is $\frac{4}{5}\Delta_o$ (Fig. 2.4(b)). The corresponding values for these ions in tetrahedral complexes are $\frac{3}{5}\Delta_t$ and $\frac{6}{5}\Delta_t$ (Fig. 2.6(b)). The CFSE values for other ions are collected in Table 2.2. This extra stability of the aquo-complexes will manifest itself in heats of hydration which will be larger than expected. If the aquo-complexes are assumed to be octahedral, then using values of Δ_o obtained spectroscopically, quantitative estimates of the CFSE can be made and subtracted from the observed heats of hydration. These 'corrected' values do indeed, along with the d^0, d^5 and d^{10} values, lie on an almost smooth curve. For further discussion of hydration energy, see G. Pass: *Ions in solutions (3): inorganic properties* (OCS 7).

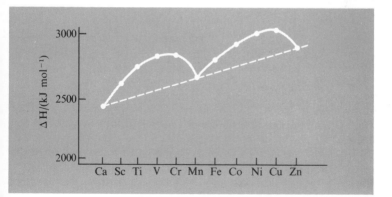

FIG. 2.7. Hydration energies for some M^{2+} ions, and the CFSE correction.

TABLE 2.2

Crystal-field stabilization energies

Number of d-electrons	High-spin octahedral	Low-spin octahedral	Tetrahedral (high-spin)[†]
0	0	—	0
1	$\frac{2}{5}\Delta_0$	—	$\frac{3}{5}\Delta_t$
2	$\frac{4}{5}\Delta_0$	—	$\frac{6}{5}\Delta_t$
3	$\frac{6}{5}\Delta_0$	—	$\frac{4}{5}\Delta_t$
4	$\frac{3}{5}\Delta_0$	$\frac{8}{5}\Delta_0$	$\frac{2}{5}\Delta_t$
5	0	$2\Delta_0$	0
6	$\frac{2}{5}\Delta_0$	$\frac{12}{5}\Delta_0$	$\frac{3}{5}\Delta_t$
7	$\frac{4}{5}\Delta_0$	$\frac{9}{5}\Delta_0$	$\frac{6}{5}\Delta_t$
8	$\frac{6}{5}\Delta_0$	—	$\frac{4}{5}\Delta_t$
9	$\frac{3}{5}\Delta_0$	—	$\frac{2}{5}\Delta_t$
10	0	—	0

† Low-spin tetrahedral arrangements are not considered because their occurrence has never been demonstrated with certainty.

Similar results are obtained with ligands other than water, and also for positive as well as dipositive ions, though the data are not so complete. This explains why, with relatively few exceptions, the corresponding stability constants of divalent ions with a given ligand fall in the 'Irving–Williams' order:

$$Mn^{2+} < Fe^{2+} < Co^{2+} < Ni^{2+} < Cu^{2+} > Zn^{2+}.$$

$$d^5 \qquad d^6 \qquad d^7 \qquad d^8 \qquad d^9 \qquad d^{10}$$

For a complex ML_x, the overall stability constant β is given by

$$\beta = \frac{[ML_x]}{[M][L]^x}$$

and is related to the heat of formation, or enthalpy $\Delta H^\ominus$ by

$$\Delta H^\ominus - T\Delta S^\ominus = -RT \ln \beta[†].$$

As it is unlikely that the entropy changes, $\Delta S^\ominus$, vary significantly along the series of metal ions, providing the ligand is unchanged, then the same variation is expected in β as occurs in $\Delta H^\ominus$.

Much early preparative work in coordination chemistry was concerned with complexes of chromium(III) (d^3), cobalt(III) (d^6 low-spin) and nickel(II) (d^8). The reason for this was that complexes of these metals are kinetically rather

† $\ln x \equiv \log_e x$, $\lg x \equiv \log_{10} x$.

stable. They do not readily dissociate and can be dissolved in many solvents without the ligands being displaced by solvent molecules. At least part of the explanation of this stability is that these ions are just those with the highest CFSE.

Somewhat less convincingly, attempts have been made to predict the relative ease of formation of octahedral and tetrahedral complexes on the basis of CFSE. Table 2.2 shows that for high-spin configurations, the largest octahedral CFSE values occur for d^3 and d^8, and the largest tetrahedral CFSE values occur for d^2 and d^7. It is true that d^3 (Cr^{3+}) and d^8 (Ni^{2+}) more readily form octahedral than tetrahedral complexes, and that d^7 (Co^{2+}) forms tetrahedral complexes more readily than do most other metal ions. However, the suggestion that tetrahedral nickel(II) could not be prepared has been proved to be quite unfounded. Comparisons between octahedral and tetrahedral CFSE's are largely vitiated by the fact that Δ_o and Δ_t are of different magnitudes and a particular stereochemistry can often be 'forced' on a metal by selecting ligands of appropriate geometry.

It must always be remembered that CFSE represent only a small fraction (perhaps 5–10 per cent) of the total energy of formation of a complex. It may be crucial in determining the energy difference in comparable cases, but can easily be swamped by comparatively small changes in the larger quantities involved.

Our concern so far has been exclusively with octahedral and tetrahedral complexes. This is not unreasonable, since these are the simplest theoretically and are certainly the best understood. However, other stereochemical arrangements are often encountered. Indeed, even in nominally 'octahedral' or 'tetrahedral' arrangements, some distortion usually occurs, strictly regular symmetry being exceptional. There are several reasons for this. Thus a complex may contain several kinds of ligand, which make their own distinctive contribution to the overall electrostatic field and lie at particular distances from the metal atom. The geometry or size of the ligands may make it impossible to attain perfect octahedral or tetrahedral symmetry. Metal ions with relatively low coordination numbers may, in the solid state, be able to increase their coordination numbers by sharing some of their ligands (especially if these are monatomic) with neighbouring metal ions. If some, but not all, of the ligands are shared, distortion is inevitable. Low coordination numbers may similarly be increased in solution by the intervention of solvent molecules. Even compounds containing identical ligands and subject to none of these effects may be distorted because of the asymmetry of the d-shell of the metal. This is known as the *Jahn–Teller effect*.

The Jahn–Teller effect

A common type of distortion is found in octahedral complexes in which the two ligands on one axis, which may be defined as the z-axis, are further away

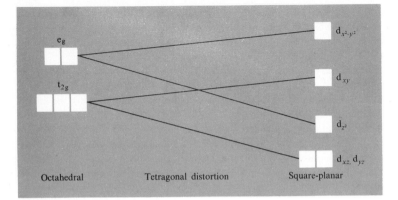

FIG. 2.8. The formation of a square-planar complex by the tetragonal distortion of an octahedral complex.

from the metal than are the other four ligands. This is a 'tetragonal' distortion and in the extreme, when the two ligands are completely removed, leads to a square-planar complex. Removal of the two ligands produces a dramatic stabilization of the d_{z^2} orbital and a somewhat smaller stabilization of the d_{xz} and d_{yz} orbitals. A secondary effect is that the loss of their neighbours on the z-axis allows the ligands in the xy-plane to move closer to the metal and so destabilize the $d_{x^2-y^2}$ and d_{xy} orbitals which however retain the same separation as in the octahedral case. This is illustrated in Fig. 2.8.

That this type of distortion is so common is due to the operation of the Jahn–Teller effect. This may be understood by looking at the formation of a complex from the point of view of the ligands. We have seen how the splitting of the d-level can be explained by the repulsion of the metal d-electrons by the ligands. However, by the same token, the ligands experience a repulsion from the d-electrons of the metal. If the d-shell is symmetrically occupied the repulsion felt by each ligand is the same, but when the d-shell is occupied unsymmetrically, the repulsions are unequal and some ligands will be prevented from approaching as close to the metal as the others. The greatest distortions are found in those configurations in which the asymmetry is in the e_g orbitals which point directly towards the ligands. Asymmetry in t_{2g} orbitals leads to much smaller distortions. If the e_g orbitals are not occupied equally, it seems reasonable that occupation of the d_{z^2} rather than the $d_{x^2-y^2}$ orbital will be preferred, since in this way only two ligands are repelled rather than four. This is the situation described in Fig. 2.8. Table 2.3 shows the expected magnitude of the effect in octahedral complexes of metal ions with various electronic configurations.

Octahedral complexes of d^8 metal ions in which the $d_{x^2-y^2}$ and d_{z^2} orbitals each contain a single electron (the other six fill the t_{2g} orbitals) are not subject

TABLE 2.3

Expected Jahn–Teller distortions in octahedral complexes

Large distortion	Moderate distortion	No distortion
	d^1	
	d^2	
		d^3
d^4 high-spin	d^4 low-spin	
	d^5 low-spin	d^5 high-spin
	d^6 high-spin	d^6 low-spin
d^7 low-spin	d^7 high-spin	
d^8 low-spin (see text)		d^8 high-spin
d^9		
		d^{10}

to a Jahn–Teller distortion. However, for ligands such as CN^-, which produce a strong crystal field, a more stable arrangement is obtained if both e_g electrons occupy the d_{z^2} orbital. A very strong Jahn–Teller effect then occurs and square-planar complexes such as the diamagnetic $[Ni(CN)_4]^{2-}$ are formed. This is very common with the divalent ions of nickel, palladium, and platinum, referred to as low-spin d^8 in Table 2.3.

The other instances of significant Jahn–Teller distortions suggested in Table 2.3 are the complexes of high-spin d^4, low-spin d^7, and d^9 metal ions,

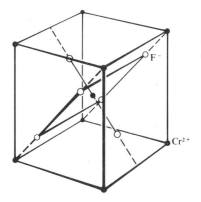

FIG. 2.9. Distorted rutile structure of CrF_2. Of the nine Cr^{2+} ions shown, eight are each shared between eight units, so there is an effective total of two Cr^{2+} ions per unit. Of the six F^- ions shown, four are each shared between two units, so there is an effective total of four F^- ions per unit. The ratio of $Cr^{2+}:F^-$ in each unit is thus $2:4$, corresponding to the stoichiometry of CrF_2.

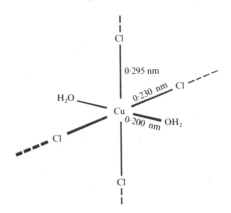

FIG. 2.10. Structure of $CuCl_2$.

e.g. Cr^{2+}, Co^{2+}, and Cu^{2+}. Very few complexes of low-spin cobalt(II) are known because of the ease with which it is oxidized to cobalt(III). However, the other two cations afford a number of authenticated examples.

Chromium(II) fluoride, CrF_2, is an ionic solid with a distorted rutile structure (Fig. 2.9). As expected with fluoride ligands, the chromium maintains a high-spin configuration ($t_{2g}^3 e_g^1$), and the electron in the d_{z^2} orbital repels the two fluoride ions on the z-axis. X-ray crystallographic measurements show that although each chromium ion is surrounded by six fluoride ions, two of the latter are at 0·243 nm and four at 0·200 nm from the metal ion. Even when applied to some covalent compounds, the theory often provides an explanation of the observed results; in copper(II) chloride dihydrate, $CuCl_2 \cdot 2H_2O$, each chloride ion is shared between two copper atoms which therefore attain a coordination number of six (Fig. 2.10). The expected distortion occurs and each copper ion is surrounded by two chlorides at 0·295 nm and two at 0·230 nm. The water molecules are closer still at 0·200 nm. This is in keeping with the fact that water produces a stronger electrostatic field than chloride (see the spectrochemical series on p. 29).

In addition to interatomic distances obtained from X-ray crystallographic measurements, confirmation of Jahn–Teller distortions is also provided by spectroscopic measurements. The further splitting of the d-levels resulting from the distortion increases the number of electronic transitions which are possible. In a distorted copper(II) complex, the separation between the t_{2g} and the d_{z^2} orbitals differs from that between the t_{2g} and the $d_{x^2-y^2}$, and two distinct transitions are possible. The absorption bands to which these give rise are rather broad and are superimposed on each other as shown in Fig. 2.11.

Unequal occupation of metal d-orbitals would be expected to distort tetrahedral complexes similarly, but since none of the d-orbitals points

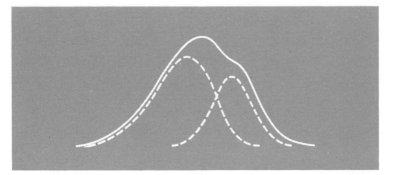

FIG. 2.11. Electronic spectrum of $[Cu(H_2O)_6]^{2+}$, showing two absorptions due to tetragonal distortion.

directly towards the ligands, the distortions are not so pronounced and are far less well documented.

The above crystal-field account of tetragonal distortions and square-planar complexes is undoubtedly superior to that provided by VB theory. Admittedly the two are compatible in the case of diamagnetic square-planar d^8 complexes. Here the $d_{x^2-y^2}$ orbital, which according to VB theory is involved in dsp^2 hybridization to accommodate the bonding electrons from the ligand, is, according to the CF account, energetically too high to be occupied by metal electrons. However, as previously mentioned, d^9 ions present VB theory with the problem of where to accommodate the extra electron. No such problem arises in the CF account. All nine metal electrons remain in the d-orbitals, and furthermore the lack of any sharp demarcation between square-planar and distorted-octahedral complexes is readily understood.

It is striking that in this outline of the successful applications of crystal-field theory, repeated references are made to transition metals of the first series but hardly any to the second and third. Oxidation states other than $+2$ and $+3$ are not mentioned at all. This is no coincidence. Oxidation states of $+2$ and $+3$ are far less common in the second and third series than in the first, and they are the oxidation states expected to be most ionic in character. It is not surprising then that a theory which eschews all possibility of covalency should find its most fruitful applications in this area; nor that attempts to apply it elsewhere are less successful. Where it is proper to apply crystal-field theory, it is a most useful and informative approach, but care must be taken to ensure that its application in a particular case is legitimate. Wherever covalency is likely, some modification of the theory will be needed. Evidence of the presence of covalency is provided by a variety of spectroscopic and magnetic techniques and it is possible to account for many of the results by introducing parameters such as the *electron delocalization factor k*. Such adjustments appreciably extend the scope of the basically electrostatic model, and are generally referred to under the heading of *ligand-field theory*.† However, when the degree of covalency is too great to be accounted for in this way, it is necessary to adopt a radically different approach. This is provided by *molecular-orbital theory*.

Molecular-orbital theory

It has already been shown how the concept of atomic orbitals can be used to describe the relative energies and spatial arrangements of electrons in an isolated atom, and how the electronic configurations of the elements can be deduced by feeding electrons into these orbitals. In the same way it is possible to define molecular orbitals (MOs) which encompass all the nuclei of a molecule and to deduce the molecular electronic configurations by feeding the available electrons into these. The restrictions that an electron will occupy the orbital of lowest energy, and that each orbital can accommodate a maximum of only two electrons, still apply.

Unfortunately, whereas the definition of atomic orbitals is possible without too much mathematical difficulty, the problem of electrons attracted by several atomic nuclei is much more difficult, and molecular orbitals can rarely be derived exactly. One of the best, and certainly the simplest approach is to start with the atomic orbitals of the relevant atoms and to derive the molecular orbitals from them by what is known as the Linear Combination of Atomic Orbitals (LCAO) method. We shall not attempt the mathematics of this process,‡ but will outline the results.

The combination of n atomic orbitals produces n molecular orbitals of which in general half have a lower energy than the original atomic orbitals (bonding molecular orbitals) and half have a higher energy (antibonding molecular orbitals). The situation is quite complicated for compounds of elements, such as the transition metals, which possess a variety of atomic orbitals. It may be helpful if a few simpler cases are dealt with first. The molecular orbitals produced by the combination of two s-orbitals, as for instance is assumed to occur when two atoms of hydrogen combine to form a molecule of H_2, are illustrated in Fig. 2.12. The separation between the bonding and antibonding orbitals depends on the extent to which the atomic orbitals overlap. That the bonding orbital is more stable than the antibonding is evident from their shapes. Electrons in the former are attracted by both nuclei whereas electrons in the latter are significantly attracted by only one nucleus.

An alternative representation, which conveniently indicates the electronic configuration of the molecule but not the shape of the orbitals, is shown for

† This name has often in the past been used synonymously with crystal-field theory. The distinction suggested here is that which is now most generally accepted.

‡ Described by C. A. Coulson in *The shape and structure of molecules* (OCS 9).

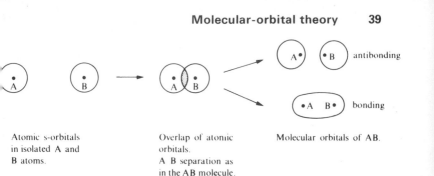

Atomic s-orbitals in isolated A and B atoms.

Overlap of atomic orbitals. A B separation as in the AB molecule.

Molecular orbitals of AB.

antibonding

bonding

FIG. 2.12. Formation of molecular orbitals by LCAO.

H_2 in Fig. 2.13. In considering the possible formation of He_2, this diagram shows that the four electrons from the two helium atoms would fill both the bonding and the antibonding molecular orbitals. As in all such cases, the increase in energy of the antibonding electrons relative to the isolated atoms is somewhat greater than the decrease in energy of the bonding electrons. He_2 is thus less stable than individual atoms of helium and cannot have a lasting existence. As with 1s orbitals, so with the 2s. Lithium ($2s^1$) forms Li_2 in the vapour phase while for beryllium ($2s^2$), Be_2 is unstable.

The combination of p-orbitals is rather more complicated as it may occur in two ways. If the two p-orbitals involved point towards each other along the internuclear axis then the bonding molecular orbital also lies along this axis. It is said to be a σ (sigma) orbital, and the bond a σ-bond (Fig. 2.14(a)). On the other hand the two combining p-orbitals may be perpendicular to the internuclear axis, in which case the bonding orbital has lobes on the opposite sides

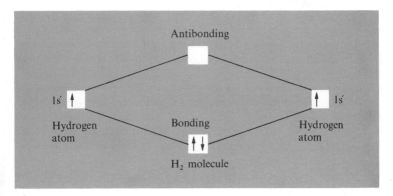

FIG. 2.13. Molecular-orbital diagram for H_2.

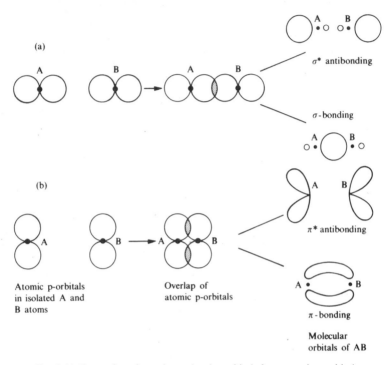

FIG. 2.14. Formation of σ and π molecular orbitals from atomic p-orbitals.

of the axes, and is said to be a π (pi) orbital (Fig. 2.14(b)). The corresponding antibonding orbitals are conventionally distinguished by an asterisk. Because of the greater overlap of the atomic orbitals, the separation of bonding and antibonding orbitals is greater for the σ-bond than for the π-bond, i.e. the σ-bond is stronger than the π-bond since it leads to greater stabilization of the bonding electrons. This is summarized in Fig. 2.15 for the O_2 molecule, though the same orbital scheme is appropriate to all diatomic molecules of elements of the first period. The formation of σ- and π-bonds is a general phenomenon, and which type is produced depends on the symmetry properties of the atomic orbitals involved. When these orbitals overlap on the axis between the two bonded atoms, a σ-bond is formed. This can occur not only for two p-orbitals but also for $s+s$, $s+p$, $s+d$, and $p+d$ orbitals. When the two atomic orbitals overlap non-axially, a π-bond is formed. Again, this can occur not only for two p-orbitals, but also for $p+d$ and $d+d$. One further point must be mentioned before turning to the more complicated situation found in transition-metal complexes. This is that for significant interaction to occur between two

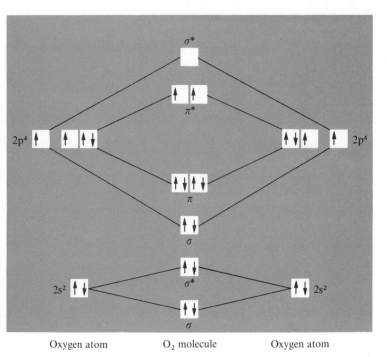

Oxygen atom O_2 molecule   Oxygen atom

FIG. 2.15. Molecular-orbital diagram for O_2. Note that the presence of two unpaired electrons satisfactorily explains the paramagnetism of oxygen.

atomic orbitals their energies must be similar. Combination of metal and ligand orbitals of grossly different energies can be neglected.

It is usual to start with an octahedral complex of the type ML_6 in which each ligand has one σ-orbital, which is commonly a p or sp^3 hybrid orbital. The possibility of π-bonding may be considered later. The only metal orbitals which can overlap with the σ-orbitals of the ligands are the e_g pair of the 3d level (assuming first-series transition elements), the 4s and the 4p. This gives rise to the MO energy-level diagram of Fig. 2.16 which can be understood if the following points are noted.

(i) The metal t_{2g} orbitals, because of their shape, do not overlap with the ligand σ-orbitals. They are 'non-bonding' orbitals and remain associated with the metal.

(ii) The shapes and positions of the molecular orbitals are not directly shown. However, of the atomic orbitals from which it is produced, a molecular orbital resembles most closely that to which it is nearest in energy.

(iii) Even when the ligands are anions (e.g. halide ions) rather than neutral molecules, it is quite immaterial whether coordination is considered to be

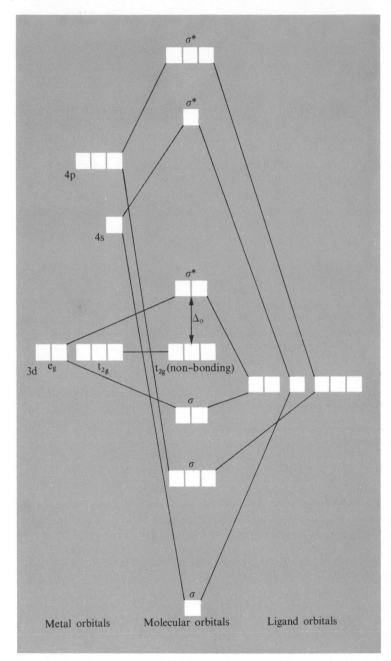

FIG. 2.16. A typical molecular-orbital diagram for an octahedral complex in which only σ-bonding occurs.

between previously formed ions or between neutral atoms. Provided that allowance is made for the appropriate transference of electrons to account for the charges on the complex ion and the ions outside the coordination sphere,† the available molecular orbitals are simply occupied by whatever electrons are available. Whether these are considered to originate from the metal or the ligands as ions or as atoms is of no importance.

(iv) A bond is purely covalent if the molecular orbitals, and hence the electrons occupying them, are shared equally by the bonded atoms. Ionic character is indicated by the extent to which these orbitals are preferentially associated with either the metal or the ligand nuclei. It follows from (ii) that the ionic or covalent character of particular electrons can be inferred from the energy of the MO they occupy, relative to the atomic orbitals from which it is produced.

The six pairs of electrons provided by the ligands are sufficient to fill exactly the low-energy bonding molecular orbitals. The electrons from the metal are then distributed between the non-bonding t_{2g} orbitals and the two σ^* orbitals immediately above them. This central portion of the diagram is exactly analogous to the crystal-field description, except that the upper pair of orbitals are no longer purely metal orbitals but are now antibonding orbitals with some ligand character. This however does not in principle alter the crystal-field description of the origin of colour in transition-metal complexes. The way in which the magnitude of Δ_o determines whether high- or low-spin configurations result and the consequent effect on magnetic properties remains the same. The superiority of MO theory lies in its flexibility in describing both extremes of ionic and covalent bonding. If the ligand atomic orbitals in Fig. 2.16 were to be drawn at a sufficiently low energy relative to the metal atomic orbitals, the bonding molecular orbitals would all be associated primarily with the ligands, while the σ^* antibonding pair as well as the non-bonding t_{2g} orbitals would remain primarily associated with the metal. The bonding described would then be predominantly ionic (i.e., electrostatic) as assumed in the CF approach. Conversely, if ligand and metal atomic orbitals were to be drawn with comparable energies, a more covalent situation would be described.

Appreciable covalency, particularly in low oxidation states, poses a rather serious problem. Metals are characteristically considered to lose electrons and form cations. However, in a complex where a single metal atom or ion is coordinated to several ligands, appreciable covalency in each bond implies a large transference of negative charge to the metal. If, for instance, a divalent metal ion were coordinated to six ligands and shared equally the six electron pairs provided by the ligands, a charge on the metal of -4 would result. The

---

† In discussing coordinate bonding, ions outside the coordination sphere are important only in that they determine the oxidation state of the metal.

problem admittedly is not so serious for metals in higher oxidation states but it is still worse for metals in zero (e.g. $Ni(CO)_4$) or even negative (e.g. $[Mn(CO)_5]^-$) oxidation states. A way out of this difficulty is provided by the concept of *back coordination*, whereby the metal not only accepts electrons from the ligands, but also donates to the ligands some of its own electrons. π-bonding provides a mechanism for this process provided that the ligands have empty orbitals of the correct shape and energy to overlap with appropriate filled metal orbitals. The key is provided by the metal t_{2g} orbitals which in Fig. 2.16 were described as 'non-bonding' only because π-bonding was specifically ignored.

With a ligand such as carbon monoxide this is quite unjustified. Its electronic configuration can be derived from Fig. 2.15 by placing ten electrons (six from oxygen, four from carbon) in the molecular orbitals. It is not necessary for the present purposes to describe in detail the shapes of all these orbitals. It is sufficient to note that there are filled σ-orbitals on both carbon and oxygen atoms and two empty π^* orbitals, as shown in Fig. 2.17. Coordination of carbon monoxide to a metal is through the carbon atom and takes place in two stages. The σ lone pair of electrons is first donated to the metal in the normal way. This is followed by π-overlap between the empty π^* orbitals and the metal t_{2g}. As the latter are invariably filled, this is equivalent to donation of electrons from the metal to the ligand.[†] This process is readily represented on an MO energy-level diagram by simply adapting the central portion of Fig. 2.16 to allow for the addition of π-bonding. As is seen in Fig. 2.18, the result is that Δ_o is larger than it would have been if only σ-bonding had occurred.[‡] In this way, MO theory provides the explanation, which CF

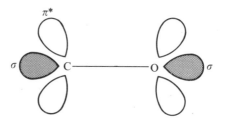

FIG. 2.17. Two occupied σ-orbitals (shaded) and an empty π^* orbital (unshaded) in CO. There is another empty π^* orbital at right angles to the one shown.

[†] This description of the bonding is oversimplified and is only intended to demonstrate how, in principle, axial donation and non-axial acceptance of electrons by the ligand is possible. In fact, to account for the various coordination numbers and stereochemistries, it is necessary to combine linearly the above orbitals of all the ligands. It is the appropriate products of this which actually overlap with the metal orbitals. In octahedral complexes there are three of these.

[‡] Although MO theory can be successfully applied to octahedral complexes, its application to other stereochemistries is more complicated and often, as a consequence, less satisfactory.

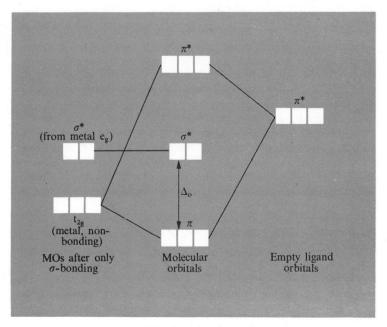

Fig. 2.18. The effect of back donation on Δ_0.

theory was quite unable to do, for the position of carbon monoxide in the spectrochemical series. Since CN^- and NO^+ are isoelectronic with CO, they too are capable of the same behaviour. Clearly, in compounds containing these ligands, the physical significance of the oxidation state assigned to the metal is inevitably rather vague.

Molecular nitrogen, or dinitrogen, N_2, is also isoelectronic with carbon monoxide, and though complexes of N_2 proved rather elusive, several have now been prepared.[†] The importance of this lies in the possibility that such compounds might lead to practicable methods of nitrogen *fixation*. It is known that in many plants the enzymes responsible for the biochemical synthesis of nitrogen-containing compounds from atmospheric nitrogen contain iron and molybdenum.

Metal carbonyls are an important class of compounds and have been of great interest, both practically and theoretically, to chemists. It may be worthwhile at this point to digress in order to outline some of their general characteristics.

[†] The first dinitrogen complex, $[Ru(NH_3)_5(N_2)]^{2+}$, was isolated in 1965, and within six years about 25 mononuclear complexes of dinitrogen had been characterized.

Carbonyls

Carbonyls are formed by almost all transition metals. Many of them are liquids; they are all hydrophobic, and most dissolve in non-polar solvents. They are, in short, typical covalent compounds. In addition, the oxidation state of the metal is invariably low (usually zero). It is clearly not surprising that crystal-field theory is unsuccessful in explaining their bonding, and that both valence-bond and molecular-orbital approaches are more appropriate. Table 2.4 lists the simpler, binary carbonyls of first-series transition metals with zero oxidation state. With the exception of $V(CO)_6$ all of these are diamagnetic. From left to right the number of metal electrons increases and the coordination number decreases. Metals with an even number of electrons form mononuclear carbonyls, those with an odd number form binuclear carbonyls, apart from $V(CO)_6$.

This is most readily rationalized by the assumption that the number of valence electrons possessed or shared by the metal (each carbon monoxide molecule contributing a pair of electrons) equals the number possessed by the next noble gas. The number of carbon monoxide molecules bonded to the metal is then simply the number needed to complete the $d^{10}s^2p^6$ configuration. This is six, five, and four for chromium, iron, and nickel respectively. Manganese and cobalt have odd numbers of electrons and make up the additional electron by dimerizing and forming metal–metal bonds (Table 2.5). In this respect, $V(CO)_6$ is peculiar. It is one of the very few carbonyls which are paramagnetic but it is significant that even here the noble-gas configuration can be attained by taking up an electron to form an anion as in $Na^+[V(CO)_6]^-$.

Besides the binary carbonyls, of which the above are the simplest examples, there are very many other mixed carbonyls, some of great industrial importance as catalysts in the production of plastics. Their study is part of the vast field of organometallic chemistry which is beyond the scope of this book. Perhaps one example will serve however to illustrate the importance of π-bonding in

TABLE 2.4

Some binary carbonyls of first-series transition metals

Number of metal valence electrons	5	6	7	8	9	10
Mononuclear carbonyls	$V(CO)_6$	$Cr(CO)_6$		$Fe(CO)_5$		$Ni(CO)_4$
Binuclear carbonyls			$[Mn(CO)_5]_2$		$[Co(CO)_4]_2$	

TABLE 2.5

Electronic configurations of binary carbonyls of Table 4

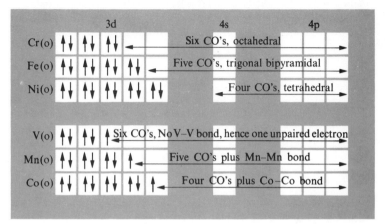

	3d					4s		4p		
Cr(o)	↑↓	↑↓	↑↓			Six CO's, octahedral				
Fe(o)	↑↓	↑↓	↑↓	↑↓		Five CO's, trigonal bipyramidal				
Ni(o)	↑↓	↑↓	↑↓	↑↓	↑↓	Four CO's, tetrahedral				
V(o)	↑↓	↑↓	↑	Six CO's, No V–V bond, hence one unpaired electron						
Mn(o)	↑↓	↑↓	↑↓	↑	Five CO's plus Mn–Mn bond					
Co(o)	↑↓	↑↓	↑↓	↑↓	↑	Four CO's plus Co–Co bond				

organometallic compounds. Zeise's salt, $PtCl_2 \cdot C_2H_4$, has been known since the early nineteenth century but its structure was not understood for over 120 years. It is now known that, in common with other unsaturated hydrocarbons, ethylene, like carbon monoxide, can be double bonded to metals by a σ- and a π-bond. In Fig. 2.19 only the relevant orbitals are drawn.

Returning to the question of bonding theory, it may be said in conclusion that the more comprehensive the theory the more limited in practice is its scope. Molecular-orbital theory is certainly the most complete and, when it can be applied conveniently, is the most satisfactory. Unfortunately it is mathematically complicated and not easily applied quantitatively. Crystal-field theory is successful if restricted to complexes of metals with the ionic oxidation states of $+2$ and $+3$, and ligands such as water and ammonia which,

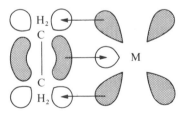

FIG. 2.19. Bonding of ethylene to a transition metal. Shaded orbitals are filled, unshaded are empty.

in MO theory, would be described as σ-donor ligands. This restriction excludes much of the chemistry of second and third transition-series metals as well as organometallic compounds. Qualitatively, valence-bond theory is applicable over a greater range of oxidation states and stereochemistries than either MO or CF theories, but it is deficient in its quantitative predictions.

PROBLEMS

2.1. Outline features of interest in three of the following compounds: (a) $Cu(NH_3)_6Cl_2$, (b) $Fe(C_5H_5)_2$, (c) $Co_2(CO)_8$, (d) $Rb_3Ni(NO_2)_5$. (*University of Essex*, modified)

2.2. The ion $Fe(H_2O)_6^{2+}$ has a magnetic moment of 4·9 B.m. while the ion $Fe(CN)_6^{4-}$ is diamagnetic. Discuss this in terms of Pauling's valence-bond approach and illustrate the usefulness and limitations of the approach.
The famous 'brown-ring test' for nitrates depends on the fact that, under the conditions of the test, nitric oxide is generated. The formula of the brown complex ion is $[Fe(H_2O)_5NO]^{2+}$ and it has a magnetic moment of 3·9 B.m. Comment. (*University of Leeds*)
(This is discussed in the text, p. 75.)

2.3. Which complex ion has the larger Δ value in each of the following pairs:
(a) FeF_6^{3-}, OsF_6^{3-} (b) $Fe(CN)_6^{3-}$, $Fe(CN)_6^{4-}$
(c) $Mo(CN)_8^{4-}$, $W(CN)_8^{4-}$ (d) $Cr(H_2O)_6^{3+}$, $Cr(H_2O)_6^{2+}$.
(The principles involved here are given in the text.)

2.4. The tetrachloronickelate(II) ion is tetrahedral, but the corresponding tetrachloroplatinate(II) ion is square-planar. Explain.

2.5. Discuss the structure of the following compounds in terms of crystal-field theory: K_3CoF_6, $[Co(NH_3)_6](ClO_4)_3$, $[Mn(H_2O)_6](NO_3)_2$, $[C_9H_8N]_2[CoCl_4]$, $[Fe(bipy)_3](SO_4)$.
Explain their magnetic properties. (*University of Bristol*)

2.6. Why has the complex ion $[Co(NH_3)_5Br]^{2+}$ an absorption maximum at about 2500 Å whereas $[Co(NH_3)_5(H_2O)]^{3+}$ has its absorption maximum below 2200 Å?

2.7. Discuss briefly three of the following:
(a) the variation in ionic radii for the ions Ca^{2+}, Ti^{2+}, V^{2+}, Cr^{2+}, Mn^{2+}, Fe^{2+}, Co^{2+}, Ni^{2+}, Cu^{2+}, Zn^{2+},
(b) the absorption spectrum of $[Ti(H_2O)_6]^{3+}$ in aqueous solution over the range 3000–7000 Å ($Å = 10^{-10}$ metre),
(c) the evidence for an inner-sphere mechanism in the reaction of Cr^{2+} with $[Co(NH_3)_5Cl]^{2+}$,
(d) the inertness to substitution of the complex $[Fe(CN)_6]^{4-}$.
(*University of Leeds*)

2.8. Write down energy-level diagrams for $[FeF_6]^{3-}$ (a high-spin d^5 system) in terms of (a) crystal-field theory, (b) molecular-orbital theory, showing both electron occupancy and ligand-field splitting parameter Δ. (*University of Sussex*)

2.9. Discuss the generalization of the crystal-field theory of the electronic structures of transition-metal complexes to the ligand-field theory. Indicate the main reasons why such a generalization is necessary, and discuss its experimental justification. (*University of Bristol*)

2.10. Give an account of the application of crystal-field theory to the explanation of the magnetic properties and the electronic spectra of tetrahedral cobalt(II) and nickel(II) complexes MX_4^{2-}. How would the explanation differ in the case of a tetrahedral complex ML_2X_2? (*University of Southampton*)

2.11. Briefly compare and contrast the application of valence-bond theory, crystal-field theory, and molecular-orbital theory to hexacoordinated complexes of transition metals. (*University of Birmingham*)

2.12. Discuss briefly the nature of the bonding in π-complexes of transition metals. Cyclopentadiene, C_5H_6, reacts with nickel carbonyl to produce a red dia-magnetic compound $NiC_{10}H_{12}$ which was originally formulated as bis(cyclo-pentadiene)nickel(0). The ^1H n.m.r. spectrum indicates however the presence of four different types of hydrogen in the ratio $5:4:2:1$, the most intense of these being found in the aromatic region. Suggest an alternative structure for $NiC_{10}H_{12}$. (*University of Sussex—modified*)

3. The chemistry of the groups

Scandium, yttrium, lanthanum, and actinium $(n-1)d^1 ns^2$

WITH the exception of actinium, which occurs as a radioactive trace impurity in uranium ores, these elements are usually found as phosphate or silicate ores. Yttrium and lanthanum invariably occur along with the other lanthanides.† Though by no means rare (apart from actinium) they have long been regarded as unfamiliar because of the difficulty experienced in separating them in a pure form.

Physical properties

	Sc	Y	La	Ac
Density/(g cm^{-3})	2·99	4·45	5·98–6·19	—
Melting point (°C)	1539	1495	920	1050
Boiling point (°C)	2727	2927	3469	—
Covalent radius/nm	0·166	0·182	0·187	—
M^{3+} crystal radius/nm	0·081	0·093	0·106	0·111
M^{3+}(aq), $E^{\ominus}$/V (see p. 96)	−2·08	−2·37	−2·52	∼ −2·6

For each element, the energies of the ns and $(n-1)d$ orbitals are very similar. All three outer electrons are easily lost and the chemistry of the elements is confined to the +3 oxidation state. Their monatomic cations are colourless and diamagnetic, and have no catalytic properties.

This is the behaviour that would be expected of main-group elements following the alkaline-earth metals. Since the main group which in fact follows the alkaline earths is the boron–aluminium group, appreciable similarity with these elements is to be expected. It is indeed found that many of the properties of the scandium group are intermediate between strongly basic calcium on the one hand, and weakly basic aluminium on the other. Yttrium and lanthanum in particular are of course very similar to the lanthanides. As is usual, electropositive character increases with atomic number and this is reflected in increasingly ionic compounds and more-basic oxides as the group is descended. Thus scandium resembles aluminium more than calcium

---

† Whether lanthanum itself is considered as one of the lanthanides or simply their precursor is merely a matter of definition and convenience.

while, as far as its properties have been studied, actinium (oxidation number apart) has more similarity with calcium.

The oxides, M_2O_3, are white solids not easily reduced and the hydroxides, $M(OH)_3$, are insoluble gelatinous materials (in the case of scandium probably best considered as a hydrated oxide). The metals themselves are very reactive, tarnishing rapidly in air and burning on heating to give the oxide and nitride. If heated, they combine directly with hydrogen, carbon, phosphorus, sulphur, and the halogens and, if powdered, they react with water with evolution of hydrogen. The hydrides, nitrides, and carbides are essentially intermediate between the ionic salt-like compounds formed by the alkaline earths, and the non-stoichiometric interstitial compounds formed by the subsequent transition elements. The hydrides for instance react with water to produce hydrogen, as do the salt-like hydrides, but are of variable composition as are the transition-metal hydrides. The metals dissolve in dilute acids producing soluble salts with strong acids (HNO_3, H_2SO_4, $HClO_4$, HCl, HBr, HI) and insoluble or sparingly soluble salts with weak acids (HF, H_2CO_3, H_3PO_4, $H_2C_2O_4$).

The significantly smaller size of the Sc^{3+} ion, compared with the other M^{3+} ions of this group, to some extent sets it apart. Its compounds are rather less ionic, and their aqueous solutions are appreciably hydrolysed. It is the only member of the group to form an alum,† the other elements being apparently too large to fit into the required crystal lattice. Its ability to form complexes, e.g. K_3ScF_6, is also more marked, though in this it still falls short of the elements in later transition-metal groups.

Titanium, zirconium, and hafnium $(n-1)d^2ns^2$

Titanium, which comprises approximately 0·6 per cent of the earth's crust, is one of the most abundant transition metals, and zirconium (0·025 per cent) is three times more plentiful than copper. Only hafnium can be considered as rare but, for a long time, difficulty in isolating the pure metals rendered them unfamiliar.

Physical properties

	Ti	Zr	Hf
Density/(g cm^{-3})	4·50	6·53	13·29
Melting point (°C)	1675	1852	2150
Boiling point (°C)	3260	3578	5400
Covalent radius/nm	0·132	0·145	0·144
M^{4+} crystal radius/nm	0·068	0·080	$\sim$0·08

† A double sulphate of mono- and tervalent cations, similar to $K_2SO_4 \cdot Al_2(SO_4)_3 \cdot 24H_2O$.

For all these elements, the stablest oxidation state is $+4$, attained by the involvement in bonding of all the outer s- and d-electrons, and producing colourless diamagnetic compounds. Though, as these facts indicate, the ns and $(n-1)$d orbitals are still close in energy, this is the first group in which the really characteristic transitional properties of variable oxidation state, colour and paramagnetism are encountered. These are seen in the $+2$ (d^2) and $+3$ (d^1) oxidation states. M^{2+}(aq) and M^{3+}(aq) ions are strong reducing agents even for titanium and, as they are still less stable for zirconium and hafnium, the number of $+2$ and $+3$ compounds of these elements is rather limited.

One of the most notable features of this group is the extremely close similarity in properties of zirconium and hafnium. As a result of the lanthanide contraction (p. 5) the covalent and crystal radii of zirconium are almost identical with those of hafnium. Because of this similarity, it was not until 1922, when spectroscopic methods were available, that hafnium was discovered, though titanium and zirconium had been known since the late eighteenth century. Hafnium has not been so extensively studied as zirconium, but it is clear that in the group oxidation state of $+4$ the only noticeable differences are physical rather than chemical. Significant differences are only to be found in complexes involving ligands of low electronegativity with the metals in low oxidation states. Only in such cases may the differing character of the 4d and 5d orbitals be expected to manifest itself.

At high temperatures, the metals are very reactive and combine directly with most non-metals. The borides, carbides, and nitrides are interstitial compounds which have high melting points and are very hard. Because of this reactivity, the pure metals are not easily obtained. The usual method is to obtain the chloride by heating the oxide with carbon and chlorine, and then to reduce this to the metal by heating with magnesium in an atmosphere of argon:

$$MO_2 \xrightarrow{\text{C/Cl}_2} MCl_4 \xrightarrow{\text{Mg}} M.$$

However, at room temperature the metals, possibly because of a protective layer of the oxide MO_2, are unreactive and indeed show a marked resistance to corrosion and chemical attack. Titanium generally dissolves in mineral acids if heated but zirconium and hafnium are unaffected whether hot or cold. Hydrofluoric acid is exceptional, particularly if additional fluoride ion is present, when complex ions such as $[MF_6]^{2-}$ are formed. Aqueous alkali has no effect. Metallic titanium has the properties of strength and hardness usually associated with transition metals and, because of its inertness and comparative lightness, has attained considerable technical importance.

Oxidation state $+4$ (d^0)

The dioxides MO_2 are white solids of very high melting point. In the form of its ore *rutile*, TiO_2 has the structure to which the ore gives its name and in which the metal ion is octahedrally surrounded by oxide ions (p. 35). ZrO_2 and HfO_2 occur with the *fluorite* structure (i.e. similar to CaF_2) when each metal ion is surrounded by eight oxide ions. All three oxides are also found in other forms. Chemically the oxides are all amphoteric, the acidity being most marked for titanium, the smallest of the three metals. Fusion with caustic alkali gives titanates, zirconates, and hafnates, $M^{IV}O_3{}^{2-}$, but in aqueous solution these are extensively hydrolysed. Acidification of titanate solutions yields salts derived from the titanyl ion, TiO^{2+}, though the independent existence of this ion is doubtful.

The halides, apart from the fluorides which are rather poorly characterized, are all known and in the vapour phase are monomeric and tetrahedral. $TiCl_4$ is a colourless, typically covalent liquid (m.p. $-23°C$), fuming strongly in moist air in which it is hydrolysed to TiO_2. $TiBr_4$ and TiI_4 are white solids (m.p. 39 and 150°C) with similar properties. The halides of zirconium and hafnium are white solids which hydrolyse less readily and less completely and evidently possess greater ionic character than their titanium analogues. This follows from the expected increase in electropositive character as the group is descended. However, even the zirconium and hafnium halides behave as Lewis acids—that is, they tend to complex with molecules or ions possessing a pair of electrons suitable for donation. As a consequence, complexes are all of the anionic variety and the aqueous chemistry, particularly of titanium(IV), involves extensively hydrolysed species. There is no evidence of simple M^{4+}(aq) ions.

Octahedral complexes of the type $[MX_6]^{2-}$ are well known. The high charge and availability of completely empty $(n-1)d$ orbitals, as well as ns and np of similar energies, make the M^{4+} ions suitable for high coordination numbers. This is particularly true for Zr^{4+} and Hf^{4+} which have the additional advantage of being larger than Ti^{4+}. $[ZrF_7]^{3-}$ and $[ZrF_8]^{4-}$ are typical.

Oxidation state $+3$ (d^1)

This has the d^1 configuration which leads to colour and paramagnetism. It is reducing in character, tending to revert to the more stable $+4$ state. Zirconium(III) and hafnium(III) reduce water and so can be prepared only in the solid state or in non-aqueous solvents. However the $+3$ is more ionic than the $+4$ state and, in the case of titanium(III) (which in the absence of a catalyst does not reduce water) aqueous solutions of violet $[Ti(H_2O)_6]^{3+}$ with a variety of anions can be prepared in which comparatively little hydrolysis occurs. Such solutions can be obtained by the reduction of aqueous titanium(IV) with zinc and acid, and are used analytically as reducing agents. Since they are readily oxidized by atmospheric oxygen they suffer from the practical disadvantage that they must be stored and used under a non-oxidizing atmosphere such as hydrogen or nitrogen.

Octahedral anionic complexes of the type $[TiX_6]^{3-}$ are readily produced along with many others involving anionic or neutral ligands.

Lower oxidation states

The M^{2+} ions are all very strongly reducing and even titanium(II) reduces water, so that it has no aqueous chemistry and only a few compounds are known.

Owing to their inability to back-coordinate (because they have no filled d-orbitals), these metals show little tendency to form complexes with carbon monoxide and similar ligands.

Vanadium, niobium, and tantalum $(n-1)d^3ns^2$ (Nb $4d^4 5s^1$)

Vanadium, comprising 0·015 per cent of the earth's crust, is fairly common; niobium and tantalum, which always occur together, are less so.

Physical properties

	V	Nb	Ta
Density/(g cm^{-3})	6·11	8·57	16·6
Melting point (°C)	1895	2470	2996
Boiling point (°C)	∼3000	4927	5425
Covalent radius/nm	0·1224	0·1342	0·1343
M^V(aq), M $E^{\ominus}/V$ (see p. 96)	−0·253	−0·65	−0·81

Each oxidation state from $+2$ to $+5$ is shown by all the metals, but as usual, it is only in the case of the first member, vanadium, that the lower states show any appreciable stability. On descending the group it becomes increasingly difficult to obtain the lower oxidation states, and it is noticeable that for vanadium the $+4$ state is ordinarily the most stable, whereas for niobium and tantalum it is the $+5$. In all cases the $+3$ and $+2$ states are strongly reducing.

Niobium and tantalum resemble each other very closely because of the lanthanide contraction, though the similarity is perhaps not quite so marked as that between zirconium and hafnium.

The metals are comparatively inert, the more so as the atomic number increases. All are stable in air but are converted to the pentoxide M_2O_5 on strong heating. They react directly with all the halogens and form interstitial compounds when heated with nitrogen or carbon. These are extremely hard and refractory. Tantalum carbide TaC has a melting point in the region of 4000°C and is used in high-speed cutting tools. The metals resist attack by acids other than hydrofluoric (which forms complex fluoro-acids) vanadium dissolving only in oxidizing acids while niobium and tantalum are unaffected even by aqua regia. Molten caustic alkalis react to form salts such as vanadates.

Apart from niobium, which at low temperatures is a superconductor, there is little call for the pure metals. The main use of vanadium is in special steels for which 'ferrovanadium', an alloy of iron and vanadium, is used. Vanadium(v) oxide is also used as a catalyst for oxidations using atmospheric oxygen, e.g. the Contact process for sulphuric acid.

Oxidation state $+5$ (d^0)

The pentoxides M_2O_5 are non-volatile, amphoteric solids, V_2O_5 being mostly acidic. It is yellowish-brown as a result of charge transfer; the other two are white.

Vanadium pentoxide dissolves in acid to produce a salt of the pale yellow vanadyl ion VO_2^+, and in strong alkali to produce the colourless vanadate ion VO_4^{3-}. Ortho-, meta- and pyrovanadates (VO_4^{3-}, VO_3^-, and $V_2O_7^{4-}$) can be obtained which are isomorphous with the corresponding phosphates and arsenates. However, whereas the interconversions are slow in the latter cases, changes within the vanadate solutions are rapid and pH-dependent. If a strongly alkaline orthovanadate solution is gradually made more acidic, the initially colourless solution turns red after the neutral point and then to pale yellow as the pH falls below 2·0. A number of complicated polymeric species are formed, and which salt separates on crystallization will depend on the relative solubilities as much as on the concentrations of the species present. Among the species present as the acidity increases are:

$$VO_4^{3-}, V_2O_7^{4-}, V_4O_{12}^{4-}, V_{10}O_{28}^{6-}, V_2O_5 \cdot nH_2O \text{ (colloidal), and } VO_2^+.$$

Niobium and tantalum pentoxides show somewhat similar, though less extreme, behaviour, and a number of niobates and tantalates have been isolated.

Because of the greater stability of the $+4$ state for the lighter element, vanadium differs from niobium and tantalum in that $V(v)_{aq}$ is a moderate oxidizing agent. The various electrode potentials (see p. 96) for vanadium are summarized in the table below (all values in volts):

$$V \xrightarrow{\sim -1\cdot2} V^{2+} \xrightarrow{-0\cdot25} V^{3+} \xrightarrow{+0\cdot36} VO^{2+} \xrightarrow{+1\cdot0} VO_2^+$$

with an overall path of $-0\cdot253$.

In heated mixtures, vanadium pentoxide liberates chlorine from concentrated hydrochloric acid and converts sulphites to sulphates.

Numerous oxyhalides of the form MOX_3 can be prepared, but only with the strongly oxidizing fluorine will vanadium form a pentahalide. All the pentahalides of niobium and tantalum are known, and the chlorides, bromides, and iodides are volatile and covalent, being dimeric in the solid state (**1**). Complex halides $[MX_6]^-$ are formed by all three metals with fluoride but not

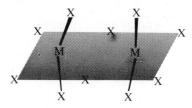

1

with the other halides. As in the previous group, higher coordination numbers are possible with the heavier elements, $[NbF_7]^{2-}$, $[TaF_7]^{2-}$, and $[TaF_8]^{3-}$ being formed. The last of these is a square antiprism.

Oxidation state $+4$ (d^1)

The chemistry of niobium and tantalum in this oxidation state is largely confined to the oxides and halides (TaF_4 is unknown). The oxides, NbO_2 and TaO_2, are obtained from the pentoxides by high-temperature reduction with hydrogen and graphite respectively. The halides are also best obtained from the pentahalides and are interesting in that only NbF_4 shows the paramagnetism expected for the d^1 configuration. The others are dimers or extended chain structures in which metal–metal bonding involves pairing of the single d-electrons on each metal atom.

By contrast, $+4$ is the most stable oxidation state of vanadium. The compounds resemble those of titanium(IV) and fall into two main groups:
(a) covalent halides,
(b) ionic oxo-salts (vanadyl) containing the blue VO^{2+} ion which in aqueous solution probably hydrates to the octahedral $[VO(H_2O)_5]^{2+}$.
The halides decompose or disproportionate on heating and are readily hydrolyzed by water.

The oxide VO_2 is a blue amphoteric solid with the rutile structure. It dissolves in strong alkali to give vanadate(IV) (or hypovanadate), VO_4^{4-} and $V_4O_9^{2-}$, species; and in non-oxidizing acids to give vanadyl salts. Solid vanadate(IV) salts with several mono- and bivalent cations are known and their solutions oxidize readily to vanadium(V).

The vanadyl salts are the most important group of vanadium(IV) compounds and give rise to a large number of complexes which are either five- or six-coordinate. $VO(acac)_2$ (2) is the best-known example of square-pyramidal coordination (acac = acetylacetonate). Coordination to the sixth, octahedral position tends to be rather weak.

2

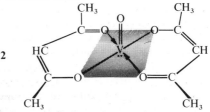

Oxidation state $+3$ (d^2)

Compounds in this oxidation state are prepared by reduction of higher oxidation states and are all reducing in character.

All the VX_3 halides are known, but niobium and tantalum give rise to a number of poorly characterized halides involving metal-atom clusters with metal–metal bonds, and in which the oxidation state and stoichiometry are often in doubt. Only vanadium has a simple solution chemistry. V_2O_3 is completely basic and dissolves in acids to give the green, paramagnetic, d^2 hexaquo-ion $[V(H_2O)_6]^{3+}$ which is only slightly hydrolysed in solution. These vanadium(III) salts are strongly reducing and indeed slowly attack water with the liberation of hydrogen. A wide variety of vanadium(III) complexes has been prepared, varying from the cationic $[V(H_2O)_6]Cl_3$ through neutral $[V(H_2O)_3F_3]$ to anionic K_3VF_6 and $K_3V(CN)_6$. For maximum stabilization, cyano-complexes require back-coordination from the metal, and it is significant that hexacyanovanadate(III) is less stable than corresponding complexes of later members of the first transition series which have more d-electrons available.

Oxidation state $+2$ (d^3)

This is the least important oxidation state even for vanadium and is very strongly reducing. Salts of the lavender-coloured d^3 ion $[V(H_2O)_6]^{2+}$ can be prepared by cathodic reduction or by the use of zinc amalgam, but decompose water fairly rapidly especially in the presence of a catalyst such as platinum. Several salts and a few complexes have been prepared.

A small number of organometallic vanadium compounds are known, and as mentioned previously, the carbonyl $V(CO)_6$ is interesting as it is the only paramagnetic carbonyl.

PROBLEMS

3.1. $E^{\ominus}$ for vanadium(V) going to vanadium(IV) is 1·0 V and hence vanadium(V) is a less strong oxidizing agent than chlorine ($E^{\ominus} = 1·36$ V). Explain therefore why warming vanadium(V) with concentrated hydrochloric acid yields chlorine almost quantitatively.

3.2. A litre of solution **A** is made up by dissolving 8g of vanadium(V) oxide in bench sulphuric acid. 25 cm³ of the solution **A** are boiled with an equal volume of 10 per cent sodium sulphite solution until no more sulphur(IV) oxide is evolved, and the resultant solution is titrated with permanganate (0·02 mol dm⁻³), 22·00 cm³ being required to oxidize the vanadium back to V(V). 5 cm³ of the original solution **A** are then shaken with zinc amalgam in a stoppered bottle until there is no further colour change, and the aqueous layer is transferred quantitatively to a conical flask containing about 50 cm³ of iron(III) sulphate solution. Titration of the resultant solution with permanganate (0·02 mol dm⁻³) required 13·20 cm³. Determine the oxidation states of the vanadium after reduction in the two cases. (V = 51, O = 16)

Chromium, molybdenum, and tungsten $(n-1)d^5ns^1$ [W $5d^4 6s^2$]

Chromium is a fairly common element; the other two are rather rarer.

Physical properties

	Cr	Mo	W
Density/(g cm^{-3})	7·19	10·22	19·3
Melting point (°C)	1890	2610	3410
Boiling point (°C)	2482	5560	5927
Covalent radius/nm	0·117	0·129	0·130

All are widely used industrially, chromium as a decorative and protective coating and in the production of stainless steels; molybdenum and tungsten in steels for the cutting edges of high-speed tools, because they remain hard and tough even at red heat. In spite of their widespread use, the chemistries of molybdenum and tungsten are imperfectly understood because of their great complexity.

The maximum possible oxidation state is $+6$, in which all the ns and $(n-1)d$ electrons are involved in bonding. All oxidation states up to this value are known, though their stabilities vary widely. Tungsten and molybdenum have the usual resemblances due to the lanthanide contraction, differences being confined largely to the lower oxidation states. They differ from chromium in that their most stable oxidation state is $+6$, while that of chromium is $+3$; they do not form monatomic cations, while chromium forms two (Cr^{2+} and Cr^{3+}); and chromium forms very few compounds in the oxidation states $+4$ and $+5$, so that there is a clear-cut distinction between its metallic behaviour in the lower oxidation states and its stereochemical resemblance to sulphur and selenium in the group oxidation state $+6$. The relevant electrode and redox potentials for acid solution are summarized in the diagram (see p. 96 for discussion of electrode potentials).

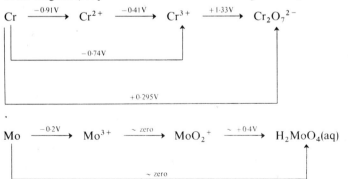

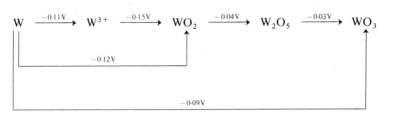

The metals have high melting and boiling points. The melting point of tungsten is the highest of any element, with the exception of carbon, which is the reason for its widespread use as a filament in incandescent lamps. They are comparatively unreactive at room temperature. Chromium dissolves in dilute non-oxidizing acids to give chromium(II) salts; molybdenum and tungsten are more resistant, though fairly concentrated nitric acid converts molybdenum to MoO_3 (concentrated nitric acid makes it passive); all react, though often only on strong heating, with oxygen, sulphur, the halogens (except molybdenum with iodine), nitrogen, and carbon. The nitrides and carbides of molybdenum and tungsten are typical interstitial compounds and, because of its extreme hardness, tungsten carbide is extensively used for tipping cutting tools. However, chromium, like the succeeding elements of the first series, has too small an atom to allow ready insertion of carbon into the interstices of the lattice and its carbide shows some properties (e.g. hydrolysis by water, producing hydrocarbons) akin to those of the ionic carbides.

All three MO_3 oxides are known and are acidic. As expected for oxides in the group oxidation state, that of the lightest element, CrO_3, is the most covalent and acidic. It is a red, crystalline solid with a chain structure, m.p. 197°C, made by the addition of concentrated sulphuric acid to a strong aqueous solution of a dichromate. It is a powerful oxidizing agent, especially towards organic substances, and is readily soluble in water giving strongly acidic solutions. MoO_3 is a white solid with a layer structure, melting at 795°C, and WO_3 is a yellow solid with a three-dimensional structure, melting at 1473°C. MoO_3 and WO_3 show no significant oxidizing properties and are insoluble in water, but dissolve in aqueous alkali giving molybdates and tungstates. The pentoxide of molybdenum and the dioxides of molybdenum and tungsten are known, but the only other really important oxide of the group, besides the trioxides, is the green chromium(III) oxide, Cr_2O_3. It is the most stable oxide of chromium and is amphoteric. The dry material is often unreactive but when precipitated from chromium(III) solutions as the hydrous oxide or 'hydroxide', it dissolves in acids to give $[Cr(H_2O)_6]^{3+}$ and in concentrated alkali to give highly hydrolysed chromate(III) species. Because of their strong reducing character, it is doubtful if pure samples of chromium(II) oxide or hydroxides have been prepared, but they are clearly basic and with non-oxidizing acids give ionic chromium(II) salts.

Oxidation state $+6$ (d^0)

Apart from oxides and oxy-anions, the only chromium(VI) compounds are oxy-halides of which chromyl chloride, CrO_2Cl_2, a deep-red unstable liquid hydrolysing rapidly in water, is the best known. There are no simple chromium(VI) halides, whereas molybdenum forms a fluoride and tungsten a fluoride, chloride, and bromide. All are volatile covalent solids made by direct action and are hydrolysed by water to the trioxide. A number of molybdenum and tungsten oxy-halides are also known.

Dissolution of the oxides in aqueous alkali produces chromates, molybdates, and tungstates which are remarkable, particularly in the case of the two heavier elements, for their tendency to polymerize. Simple chromates are isomorphous with sulphates and selenates but are highly coloured owing to charge transfer between the metal and oxygen. Increasing acidity causes condensation to dichromates:

$$2CrO_4^{2-} + 2H^+ \rightleftharpoons Cr_2O_7^{2-} + H_2O$$

and even to tri- and tetrachromates ($Cr_4O_{13}^{2-}$). The oxidizing properties of the acid solutions may be judged from the oxidation potential $E^\ominus(Cr_2O_7^{2-}, Cr^{3+}) = +1\cdot33$ V, but are much weaker in alkaline solution. In acid solution dichromate is widely used as an inorganic oxidizing agent.

Simple MoO_4^{2-} and WO_4^{2-} ions occur only in alkaline solution. On being made weakly acidic, polymerization into complicated polymolybdate(VI) and polytungstate(VI) ions occurs and hydrated salts of the type $(NH_4)_6Mo_7O_{24}$ and $K_6H_2W_{12}O_{40}$ etc. can be separated. Further acidification gives colloidal monohydrated molybdic and tungstic acids, $H_2MoO_4 \cdot H_2O$ which even in the presence of water convert on warming to the anhydrous acids H_2MoO_4. The above polymerized anions are said to be derived from *isopoly* acids, being based on only *one* acid. *Heteropoly* acids are based on a second oxy-acid besides molybdic or tungstic, and give rise to a further large class of compounds. The additional hetero atom may vary widely, but is commonly boron, silicon, phosphorus, arsenic, or another transition metal. The general formula is $H_wA_xM_yO_z$, where A is the hetero atom and M is Mo or W. Ratios of $y/x = 6$ to 12 inclusive are known, and are distinguished as 6-acids, 12-acids, etc. A typical salt is $(NH_4)_3[PMo_{12}O_{40}]$ precipitated in the familiar test for phosphates.

In spite of the obvious structural complexity, it appears that both iso- and heteropoly anions are based on octahedra of oxygen atoms with molybdenum or tungsten at their centres. The total structures are then built up from a number of these octahedra by sharing corners or edges, but not faces. In the heteropoly anions the hetero atoms lie at the centre of octahedral or tetrahedral holes surrounded by MO_6 octahedra. This is quite analogous to the situation found in the preceding group though, for reasons which are not obvious, the phenomenon is more extensive here.

Reduction of molybdates and tungstates produces rather surprising results. A variety of mild reducing agents converts acidified solutions into molybdenum and tungsten 'blues'. These are colloidal solutions of oxides probably in oxidation states $+6$ and $+5$, but their composition is not clear. In the case of solid tungstates of alkali metals, high-temperature reduction with hydrogen produces chemically inert, electrically conducting materials with a bronze sheen—*tungsten bronzes*. They are non-stoichiometric with compositions approaching $M_x^I WO_3$ ($x < 1$) and contain tungsten(VI) and tungsten(V). The metallic properties are apparently due to the presence of d-electrons (associated with tungsten(V)) which are delocalized throughout the lattice as in metals.

Chromate, molybdate, and tungstate solutions react with hydrogen peroxide to give rather unstable peroxo-derivatives. The most familiar of these is that produced by the reaction used for the qualitative detection of dichromates or hydrogen peroxide. Acidified solutions of the two give the blue peroxide, CrO_5, which rapidly decomposes with liberation of oxygen into green Cr^{3+}(aq). Stabilization may be effected by extraction into ether or by complexing with donor molecules such as pyridine. The products are virtually diamagnetic and are therefore presumed to contain chromium(VI), the structure of the pyridine complex being **3**. In alkaline solutions, paramagnetic MO_8^{3-} species containing M(V) are produced.

3

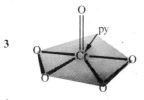

Oxidation state $+5$ (d^1)

There are only a few compounds in this oxidation state known for any of the elements, and in the case of molybdenum and tungsten they contrast with the corresponding $+6$ state compounds in being more generally coloured, because there is a d-electron which can be excited. The ones that have been prepared are usually paramagnetic and include the fluorides of chromium and molybdenum, chlorides of molybdenum and tungsten, and tungsten(V) bromide, all of which are covalent and hydrolysed by water.

Molybdenum(V) oxide is known, as are the MO_8^{3-} peroxo-compounds mentioned above. A number of complexes with high coordination number have been obtained, for example $[Mo(CN)_8]^{3-}$ by permanganate oxidation of the more stable $[Mo(CN)_8]^{4-}$.

Oxidation state $+4$ (d^2)

This too is rare. Chromium(IV) fluoride and all three chlorides occur and are covalent and hydrolysed by water. The oxides MO_2 are known and have the rutile structure. The most stable complexes are the yellow octacyanides of molybdenum and tungsten.

Oxidation state +3   (d³)

This is much the most stable oxidation state for chromium, contrasting strongly with molybdenum and tungsten. Indeed, tungsten(III) is discernibly less stable than molybdenum(III)—an example of the general trend that lower oxidation states become less stable on descending a group.

Chromium(III) compounds are readily obtained by reduction of chromium(VI) in acid, while the reverse oxidation may be effected by bromates in acid or hydrogen peroxide in alkaline solutions. The redox properties of chromium(III) form a steady trend with those of its two neighbours in the first series. Vanadium(III) is strongly reducing, chromium(III) has no significant oxidizing or reducing tendencies, while iron(III) is a moderate oxidizing agent.

Chromium(III) forms an enormous number of complexes, particularly when nitrogen is the donor atom, virtually all being six-coordinate and octahedral. Spin-pairing is not possible and most of these have magnetic moments equal to the spin-only value (3·87 B.m.) expected for three unpaired electrons. The importance of these compounds in early work in coordination chemistry was largely due to their kinetic inertness. This arises because in octahedral complexes the three d-electrons half fill the t_{2g} level giving a spherical and therefore stable cation. $[Cr(H_2O)_6]^{3+}$ is violet, but replacement of H_2O by other ligands usually produces green compounds. The range of complexes includes cations such as $[Cr(NH_3)_6]^{3+}$, anions such as $[Cr(CN)_6]^{3-}$, and species such as trioxalatochromate(III). The last of these is typical of octahedral complexes formed from bidentate ligands in that it exists in two non-superimposable mirror-image forms, **4a** and **4b**, where each loop represents oxalate.

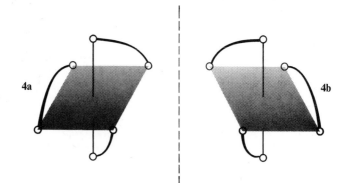

Molybdenum(III) chloride and bromide and $[MoX_6]^{3-}$ complexes can be obtained, but no such compounds of tungsten are known. One of the more

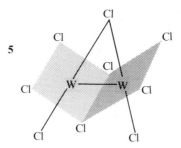

5

stable tungsten(III) complexes is $[W_2Cl_9]^{3-}$ (5) which is diamagnetic with strong tungsten–tungsten bonding. $[Cr_2Cl_9]^{3-}$ has a similar structure but is paramagnetic and shows no evidence of chromium–chromium bonding.

Oxidation state +2 (d^4)

Chromium(II) is strongly reducing (more so than iron(II)) but in the absence of catalysts is just stable in the presence of water. Aqueous solutions contain the pale-blue cation $[Cr(H_2O)_6]^{2+}$ and are used to remove oxygen from other gases. Work on Cr(II) must therefore be carried out in an inert atmosphere. There is a well-defined series of high-spin compounds (magnetic moment about 4·9 B.m.) in which is found the Jahn–Teller distortion expected from the $t_{2g}^3 e_g^1$ configuration. The d^4 configuration is the first case so far in which spin-pairing in octahedral compounds is possible. A number of low-spin complexes are known, with ligands such as cyanide and bipyridyl, with only two unpaired electrons (the t_{2g}^4 configuration). One of the more interesting compounds is the binuclear acetate, $Cr_2(CH_3CO_2)_4·2H_2O$ (6) which is almost diamagnetic due to interaction between the two metal atoms. Note the close similarity with copper(II) acetate (p. 90).

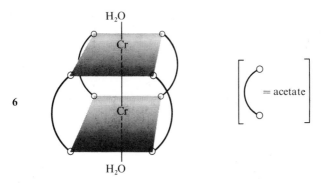

6

The most striking feature of molybdenum(II) and tungsten(II) chemistry is the absence of simple compounds, and their propensity to form metal–metal bonds, which leads to the stability of 'cluster' compounds. Thus the 'dichlorides' are best formulated as $[M_6Cl_8]^{4+}4Cl^-$ where the cation consists of an octahedron of metal ions with chloride ions attached to the eight faces. In the diagram of the cation, **7**, only one of the chlorine atoms is shown.

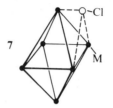

The strong tendency to metal–metal bond formation in this part of the periodic table is reflected in the very high melting and boiling points of the metals. The persistence of these clusters is as if, in the reaction of chlorine with the metal, the chlorine had been unable to break down the M–M bonding any further than the octahedral cluster.

The chemistry of lower oxidation states is confined to complexes with π-bonding ligands, such as carbonyls, of which $M(CO)_6$ are best known. They are diamagnetic, monomeric, octahedral molecules giving colourless volatile solids. A less common example of π-bonding occurs in dark brown solid dibenzene chromium **8**. This is a so-called *sandwich* compound, the metal being situated between the two parallel benzene rings. The bonding is similar to that in ferrocene (p. 76), with which dibenzene chromium is isoelectronic.

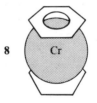

PROBLEMS

3.3. Explain why dichromate, but not permanganate, may be used volumetrically in acid solution in the presence of chloride ions.

3.4. Devise a test for chloride in the presence of bromide and/or iodide which uses the formation of a chromium compound.

Manganese, Technetium, and rhenium $(n-1)d^5 ns^2$

Manganese is quite abundant (nearly 0·1 per cent of the earth's crust) and is found largely in the form of its oxides. Rhenium, discovered only in 1925, is rare, and technetium, as its name implies, is artificial, all its isotopes being short-lived. It was first characterized as a decay product in the bombardment of molybdenum with neutrons.

Physical properties

	Mn	Tc	Re
Density/(g cm^{-3})	7·21	11·5	21·02
Melting point (°C)	1244	2200	3180
Boiling point (°C)	2097	...	5627
Covalent radius/nm	0·117	0·127	0·128

The relevant electrode and redox potentials for manganese are:

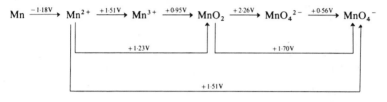

Manganese is widely used in the steel industry as a scavenger and for its toughening characteristics; rhenium has some uses as a catalyst, but the transient nature of technetium precludes any extensive use. All three metals show oxidation states up to +7 but, as expected, technetium and rhenium are more stable in the higher ones and manganese in the lower. Manganese(VII) is strongly oxidizing but manganese(II) has an extensive cationic chemistry, whereas technetium and rhenium have none at all. Manganese forms only the very rapidly hydrolysed MnF_3 and $MnCl_3$ and the ionic divalent halides; technetium and rhenium form a number of relatively volatile halides in oxidation states +6 and +4 for technetium, +7 and +6 for rhenium.

All the M_2O_7 oxides are known and are acidic. However, while Tc_2O_7 and Re_2O_7 are only moderately oxidizing, Mn_2O_7 decomposes explosively to manganese(IV) oxide. Its preparation as a dark green oil when potassium permanganate is added to concentrated sulphuric acid is best avoided! Other oxides of manganese are MnO_2, which dissolves in acids on heating and acts as an oxidizing agent giving manganese(II) solutions; Mn_3O_4 which is the most stable; and MnO which is almost entirely basic and, when precipitated as the hydroxide, oxidizes rapidly in air to Mn_2O_3. Mn_3O_4 occurs as hausmannite and is formed when any oxide of manganese is heated above 940°C. MnO, Mn_3O_4, and Mn_2O_3 are simply related. The oxide ions are close-packed and the manganese ions occupy some or all of the 'holes' between them. When all the holes are occupied MnO results, when three-quarters are occupied, Mn_3O_4 results, and when two-thirds are occupied,

Mn_2O_3. Obviously appreciable non-stoichiometry is possible. The oxides of technetium and rhenium other than M_2O_7 may be made by heating M_2O_7 with the metal but are unstable below oxidation state $+4$.

Oxidation state $+7$ (d^0)

This is the most stable oxidation state for technetium and rhenium but only occurs for manganese compounds containing oxygen: Mn_2O_7, MnO_4^- and MnO_3F. All the M_2O_7 oxides dissolves in water to give acidic solutions of HMO_4. As has already been mentioned, the intense colour of the MnO_4^- ion is a reflection of the greater oxidizing power of manganese(VII) compared to technetium(VII) and rhenium(VII)(see p. 14).

Permanganate is a well-known volumetric oxidizing agent, but its ability slowly to oxidize water ($E^\ominus = 1\cdot23$ V) prevents its being a primary standard, unlike dichromate. (See Appendix for discussion of electrode potentials.) Furthermore, the oxidizing power and the behaviour of permanganate depend upon the pH. In acid solution it is reduced to Mn^{2+}(aq) according to the half-equation:

$$MnO_4^- + 8H^+ + 5e^- = Mn^{2+} + 4H_2O \quad (E^\ominus = 1\cdot51 \text{ V})$$

with the corresponding electrode potential being given by

$$E/V = 1\cdot51 + \frac{0\cdot059}{5} \lg \frac{[MnO_4^-][H^+]^8}{[Mn^{2+}]}.$$

In neutral or alkaline solution the product is hydrated manganese(IV) oxide:

$$MnO_4^- + 3H_2O + 3e^- = MnO(OH)_2 + 4OH^- \quad (E^\ominus = 1\cdot695 \text{ V}).$$

There is another reaction possible in fairly strong alkali, producing manganate(VI):

$$MnO_4^- + e^- = MnO_4^{2-} \quad (E^\ominus = 0\cdot564 \text{ V})$$

but this reaction is not really important except in the presence of barium ions when the manganate(VI) is 'trapped' as the very insoluble barium salt.

Technetium and rhenium form oxo-halides with fluorine, chlorine, and even, in the case of rhenium, with bromine, but the yellow solid ReF_7 is the only simple halide. The black sulphides Tc_2S_7 and Re_2S_7 are known. They are readily decomposed by heat to MS_2 and sulphur.

Oxidation state $+6$ (d^1)

This is not a particularly common oxidation state for any of the elements. This may appear rather surprising for technetium and rhenium but is probably a reflection, not so much on any inherent instability, as on the greater stability of the $+7$ and $+4$ oxidation states into which their $+6$ state compounds tend to disproportionate.

Manganese(VI) is known only as the dark green MnO_4^{2-} ion which is obtained by fusing manganese dioxide with potassium hydroxide and an oxidizing agent such as potassium nitrate. It is stable only in very basic solutions; in neutral or acidic solutions it disproportionates into MnO_4^- and MnO_2. Besides the MO_4^{2-} salts and the oxides, technetium and rhenium also form fluorides and chlorides, which are yellow and green rather volatile solids, and a few oxo-halides.

Oxidation state $+5$ (d^2)

Sodium manganate(V) prepared by careful reduction of MnO_4^{2-} is one of the very few authenticated examples of manganese(V). Technetium(V) and rhenium(V) are only slightly more common, their compounds being confined to some halides, oxo-halides, and a few complexes. Again, disproportionation into $+7$ and $+4$ states appears to be the reason for the paucity of M(V) compounds.

Oxidation state $+4$ (d^3)

This is a stable oxidation state for technetium and rhenium and appears to be more stable for manganese than might be expected. The stability of manganese(IV) is largely achieved either by insolubility or by complexing. MnO_2 is a black solid, virtually insoluble in water. Strong heating in air converts it to Mn_2O_3 or Mn_3O_4, and reducing agents readily convert it to manganese(II). The water-soluble compounds of manganese(IV) are largely confined to the fluoro- and chloro-complexes $[MnX_6]^{2-}$. Their stability may well be due to the symmetrical t_{2g}^3 configuration, but the oxidizing tendency of manganese(IV) is too great to allow complexing with bromide or iodide.

In contrast, technetium and rhenium in this oxidation state have practically no oxidizing properties and given rise to oxides, sulphides, and halides. Complexes of the form $[MX_6]^{2-}$ are known for all combinations of Tc or Re and halogen except $[TcF_6]^{2-}$.

Oxidation state $+3$ (d^4)

Manganese(III) is strongly oxidizing and has a tendency to disproportionate into MnO_2 and Mn^{2+}. It is stabilized by oxygen (Mn_2O_3) and fluorine and by complexing, but the simple cation is unstable in water which it oxidizes. The only manganese(III) halide stable at room temperature is the ruby-red fluoride which decomposes on heating to MnF_2 and fluorine. $MnCl_3$ decomposes similarly but below room temperature. The range of complexes is not very extensive, being confined largely to $[MnX_5(H_2O)]^{2-}$ (X = F and Cl), $[Mn(CN)_6]^{3-}$, and compounds with oxygen-donor ligands such as acetylacetonate. Apart from the cyanide, these have the high-spin $t_{2g}^3e_g^1$ configuration. The cyanide is low-spin, t_{2g}^4.

Little is known of technetium(III), but rhenium(III) is readily oxidized to rhenium(IV) and (VII), unless stabilized by M–M bonding as in the halides and their anionic complexes. The halides of rhenium are the reverse of those of manganese in that they do not include the fluoride but are found only with the less oxidizing halogens, chlorine, bromine, and iodine. They are trimeric with three Re–Re bonds (**9**). Dissolution of the halides in the hydrohalic acids

9

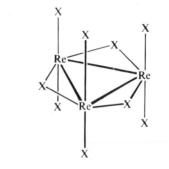

produces complex anions, formally $[ReX_4]^-$. These are diamagnetic and were originally thought to be tetrahedral and low-spin, but it has been shown that the Re_3X_9 unit persists in solution and they are correctly formulated as $[Re_3X_{12}]^{3-}$. The three extra X^- ions are attached equatorially to the three rhenium atoms in **9**, and the Re–Re bonding accounts for the diamagnetism.

Oxidation state $+2$ (d^5)

The stabilizing effect of the d^5 configuration is well illustrated by manganese(II) which is much the most stable oxidation state for this metal. Mn^{2+} has an extensive aqueous chemistry and all its simple salts are ionic. Only when the solutions are made alkaline, and the covalent hydroxide $Mn(OH)_2$ is formed, does it oxidize easily. For technetium and rhenium however, the effect of the d^5 configuration is more than offset by the generally reduced stability of their lower oxidation states, and their M(II) chemistry is confined to complexes with ligands with donor atoms such as arsenic, well known for their ability to stabilize low oxidation states by means of π-bonding.

It is notable that though Mn^{2+} is stable against oxidation, its complexes readily dissociate—their stability constants are low and they are kinetically labile. The absence of any crystal-field stabilization in high-spin octahedral complexes of d^5 ions has already been discussed, and the greater size of the Mn^{2+} ion as compared to its first-series neighbours is a further contributing factor. Low-spin octahedral d^5 ions have large crystal-field stabilization energies, but spin-pairing is achieved for manganese(II) only by ligands high in the spectrochemical series such as CN^-. As these involve π-bonding, the introduction of covalency brings with it a susceptibility to oxidation. High-spin octahedral coordination is found with many chelating agents and some $[MnX_6]^{4-}$ complexes as well as the cyanides $[Mn(CN)_6]^{4-}$ are known. Four-coordinate complexes of the type $[MnX_4]^{2-}$ can be obtained if non-aqueous solvents such as ethanol are used. These are tetrahedral, though there is some tendency to attain octahedral coordination by polymerization involving halide bridges.

Although, for manganese, the separation of the 4s and 3d orbitals is sufficient to make Mn^{2+}(aq) (loss of s^2) significantly more stable than Mn^{3+}(aq) (loss of d^1s^2), the d-orbitals of elements in the mid-region of the transition series have not yet sunk energetically into the inert electron core (Fig. 0.3, p. 5). Back donation of electrons in these orbitals is therefore relatively easy and, since there are several such electrons, stabilization of a number of low oxidation states by π-bonding is to be expected. This is indeed the case, and for manganese oxidation states as low as -3 have been reported, though this reflects the formalism involved in the concept of oxidation state rather than the actual distribution of charge.

A large number of the compounds in low oxidation states are carbonyls or their derivatives. Amongst the more interesting are the binuclear carbonyls $[M(CO)_5]_2$ (M = Mn, Tc, and Re) which are binuclear because of the odd number of electrons on the metal atoms and the 'necessity' of attaining the noble-gas configuration (see Table 2.5, p. 47). The interesting feature of these is that there are no bridging carbonyl groups—the two portions of the molecule are held together solely by the M–M bond (**10**).

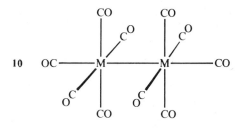

10

PROBLEMS

3.5. For permanganate, the redox potential E in volts is given by

$$E = 1.51 + \frac{0.059}{5} \lg \frac{[MnO_4^-][H^+]^8}{[Mn^{2+}]}.$$

Assuming that $[MnO_4^-] = [Mn^{2+}]$, calculate the redox potential at pH 0, 2, 4, and 6 and hence explain why the ability of permanganate to oxidize halide ions to free halogen depends upon the acidity of the solution.

3.6. 'Manganese is the most versatile element of the periodic table.' Discuss this statement and show to what extent manganese may be considered a typical transition metal. (*University of Essex*)

3.7. Manganese(III) fluoride has a distorted rutile structure in which each manganese ion is surrounded octahedrally by six fluoride ions and each fluoride ion is linearly between two manganese ions. Discuss the distortion which might be expected in the structure.

Iron, ruthenium, and osmium $(n-1)d^6 ns^2$ [Ru $4d^7 5s^1$]

After aluminium, iron is the most abundant metal, comprising over 5 per cent of the earth's crust and occurring in the form of oxide, sulphide, and carbonate ores. Ruthenium and osmium are exceedingly rare and, along with the other *platinum metals*,† occur in small amounts in copper and nickel ores from which they are frequently obtained as by-products.

Physical properties

	Fe	Ru	Os
Density/(g cm^{-3})	7·87	12·41	22·57
Melting point (°C)	1535	2250	3000
Boiling point (°C)	3000	3900	5000
Covalent radius/nm	0·116	0·124	0·126

The extraction of iron from its ores is a complex and still imperfectly understood process. The essential reduction of the ore (smelting) is effected in a blast furnace. A mixture of the ore with limestone and coke is introduced into the furnace and combustion of the coke is facilitated by air blown in at the bottom. The carbon monoxide produced by the burning coke reduces the ore, and impure 'pig iron', containing sulphur, phosphorus, silicon, carbon, and other metals as impurities is tapped continuously from the bottom of the furnace. Silicateous materials are converted by the limestone to a slag which is also continuously removed from the top of the molten metal. The impurities in the pig iron are then removed by oxidation in the presence of a suitable 'flux' which depends on the particular impurities (silica for basic oxides, limestone for acidic oxides). A slag is removed, and alloying metals added to produce the required steel.

Ruthenium and osmium find some uses as catalysts, as do the other platinum metals, and osmium(VIII) oxide is used as a biological stain and as an oxidizing agent for alkenes. The enormous number of uses of iron and steel and their fundamental importance in modern civilization need no elaboration.

† In the later parts of the transition series, as the $(n-1)$d orbitals become entrenched in the inert electron core (Fig. 0.3), the higher oxidation states (particularly the group oxidation states) become less important, and 'horizontal' rather than 'vertical' similarities are increasingly obvious. For this reason it is often useful to consider iron, cobalt, and nickel separately from ruthenium, rhodium, palladium, and osmium, iridium, platinum. These last six elements are known collectively as the platinum metals.

Iron is the first element in the first transition series which does not show its group oxidation state. Its highest oxidation state is $+6$, and even this is unstable and powerfully oxidizing, whereas it has an extensive cationic chemistry in the $+2$ and $+3$ states. This contrasts with the behaviour of ruthenium and osmium for which the higher oxidation states have a markedly

$$Fe \xrightarrow{\ -0.44V\ } Fe^{2+} \xrightarrow{\ +0.77V\ } Fe^{3+}$$

greater stability. In passing from ruthenium to osmium the effect of the lanthanide contraction is still evident, but to a diminishing extent, and a number of differences consequent on descending the group become apparent. The group oxidation state is known for both, but compounds of ruthenium (VIII) are strongly oxidizing and few in number, while osmium(VIII) is a much weaker oxidizing agent and its compounds more numerous. The most stable state for ruthenium is $+3$ but for osmium is $+4$; lower states are strongly reducing.

Pure iron is soft, malleable, and weldable, but its physical characteristics may be profoundly altered by the addition of other metals, or of non-metals, even in minute amounts. Ruthenium and osmium are hard and brittle, and osmium is the densest element known. None of the metals react with oxygen-free water, but iron decomposes steam reversibly at red heat. To a large extent the differing reactivities of the metals can be rationalized on the basis of their differing susceptibility to oxidation. Osmium is oxidized easily to the tetroxide; ruthenium with more difficulty, and only the strongest oxidizing agents raise the oxidation state of iron above $+3$. Iron dissolves readily in dilute acids, but oxidizing acids, such as concentrated nitric acid, form an impervious layer of oxides (Fe_3O_4 and Fe_2O_3) thereby rendering it passive. Ruthenium is not even attacked by aqua regia, but reacts violently with concentrated hydrochloric or nitric acids if excess potassium chlorate(V) is added. Osmium is also resistant to attack by non-oxidizing acids, but is oxidized to the tetroxide by concentrated nitric acid. Fusion with caustic alkali in air gives ruthenates(VI) and osmates(VI), but the iron compounds are not formed under comparable conditions. All combine with oxygen on heating, giving Fe_3O_4 and Fe_2O_3 (depending upon conditions), RuO_2, and OsO_4.

Three oxides of iron are known, the third being FeO. Each tends to be non-stoichiometric, and interconversion by oxidation or reduction is easy. The reason is that the structure of each is essentially based on a cubic close-packed lattice of oxide ions with metal ions in the interstices. In FeO, all the metal ions are Fe^{2+} and in Fe_2O_3 all are Fe^{3+}. Fe_3O_4 is intermediate, containing one third Fe^{2+} and two-thirds Fe^{3+}. Though different interstitial sites are occupied by the metal ions, conversion of one oxide to another is possible without any major alteration of the structure.

Ruthenium and osmium each have two oxides, MO_2 and MO_4. The tetroxides are volatile, poisonous solids. Both are oxidizing agents, but of the two, RuO_4 is the stronger, exploding violently with organic materials such as alcohol or acetone. They liberate chlorine from hydrochloric acid but, in the case of osmium tetroxide, only if the acid is concentrated. When dissolved in aqueous alkali, RuO_4 is reduced to a 'per'-ruthenate(VII) (RuO_4^-) and then to a ruthenate(VI) (RuO_4^{2-}), but osmium tetroxide hydrolyses without reduction to give $[OsO_4(OH)_2]^{2-}$ and requires mild reduction to give an osmate(VI), which in contrast to a ruthenate is of the form $[OsO_2(OH)_4]^{2-}$. Their differing oxidizing ability is also shown in their respective ease of preparation. Osmium tetroxide is obtained by heating the metal in air or by the oxidation of osmium in a lower oxidation state with concentrated nitric acid, whereas ruthenium tetroxide requires much stronger oxidizing conditions such as alkaline fusion with permanganate, followed by acidification and distillation. Both tetroxides are tetrahedral molecules, and dissolve in non-polar solvents such as benzene or carbon tetrachloride, when their dipole moments are practically zero.

The dioxides are dark, almost black, solids with the rutile structure. They can be reduced by hydrogen to the metal but only in the case of the osmium compound does heating in air produce the tetroxide.

Oxidation state +8   (d^0)

This is unknown for iron, and for ruthenium and osmium is confined to the tetroxides and complexes such as $[OsO_4F_2]^{2-}$ derived from the oxo-anions mentioned above. It is interesting to note that this highest oxidation state can be achieved by combination with oxygen, but not with fluorine alone.

Oxidation state +7   (d^1)

Per-ruthenates are the only examples of this state and are best prepared by oxidation of RuO_4^{2-} with chlorine, stopping before RuO_4 is produced. The dark-green aqueous solutions are unstable, decomposing into RuO_4^{2-} and RuO_2. The per-ruthenate ion is tetrahedral like the MO_4^- ions of the previous group, but the salts are not isomorphous.

Oxidation state +6   (d^2)

All three form oxo-anions in this oxidation state, and ruthenium and osmium also form volatile, rather unstable hexafluorides. The formation of FeO_4^{2-} (ferrates(VI)) requires very strong oxidizing conditions such as the action of fused KOH and KNO_3 on iron filings. The ferrate(VI) ion is tetrahedral, and the salts are isomorphous with sulphates, chromates, and manganates. In aqueous solution it is a stronger oxidizing agent than permanganate. It oxidizes ammonia to nitrogen in the cold, and acidification immediately causes liberation of oxygen from water.

The production of ruthenates(VI) and osmates(VI) from alkaline solutions of the tetroxides has already been mentioned. The relative ease of this reduction for the ruthenate and the relative difficulty in oxidizing it are indicative of its greater stability as compared to the osmates. Like the ferrate ion, RuO_4^{2-} is tetrahedral, but the osmate ion in contrast is octahedral (**11**). Replacement of the hydroxyl ions by chloride, cyanide, oxalate, etc., gives a series of *osmyl* complexes.

11

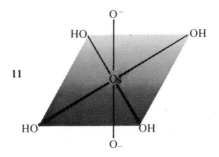

Oxidation state +5   (d^3)

This is poorly characterized, being unknown for iron, and represented for the other two metals solely by the fluorides and derived complexes. The pentafluorides are green solids, and like the fluoro-complexes are readily hydrolyzed by water. RuF_5 is obtained by direct fluorination of ruthenium, and OsF_5 by reduction of OsF_6 with iodine in iodine pentafluoride.

Oxidation state +4   (d^4)

Iron(IV) occurs in mixed oxides such as Ba_2FeO_4 and a few complexes with ligands containing polarizable arsenic or phosphorus donor atoms. It is far more stable for ruthenium and in particular osmium, there being more osmium compounds in this oxidation state than in any other. Besides the oxides there is a fluoride of ruthenium(IV) and all the tetrahalides of osmium except the iodide.

All the complexes of ruthenium(IV) and osmium(IV) appear to be octahedral and low-spin. They include an extensive range of halide complexes of the types $[MX_6]^{2-}$ and $[MX_5(OH)]^{2-}$ of which those of ruthenium are rather more easily hydrolysed in aqueous solution than are those of osmium.

Oxidation state +3   (d^5)

This is the highest of the common oxidation states of iron and is the most stable state of ruthenium. For osmium it is much less stable. In an intermediate oxidation state a sequence of this type, with the second member of the group having the most stable compounds, is not surprising. Indeed iron(III), in its high-spin compounds at least, would no doubt be less stable with respect to its lower oxidation states were it not for the presence of the symmetrical half-filled d-shell.

Many iron(III) compounds closely resemble those of chromium(III) and aluminium(III). Iron(III) halides are formed with all the halogens except iodine. The mild oxidizing properties of Fe^{3+} prevents the formation of iron(III) iodide—compare Cu(II)—and iron(III) bromide is thermally unstable. In the vapour and in non-donor solvents the chloride **12** is dimeric, each iron atom being tetrahedrally coordinated (compare Al_2Cl_6).

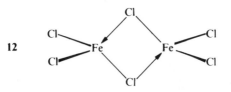

Though iron(III) oxide has no acidic properties (the so-called *ferrites*, $M^I FeO_2$, obtained by the action of alkali-metal hydroxides, are actually mixed oxides and do not contain discrete FeO_2^- ions), it is only a weak base with the result that iron(III) salts are appreciably covalent and their aqueous solutions extensively hydrolysed. The hydrated ion $[Fe(H_2O)_6]^{3+}$ is amethyst-coloured and is found in iron(III) alums. However in aqueous solutions, unless they are strongly acidic, hydrolysis occurs, protons are released, and several basic species are formed which are yellow, owing to charge-transfer transitions. The small size and high charge of the Fe^{3+} ion are responsible for this behaviour and also for the ease with which it forms a wide variety of complexes. A familiar example of this, used as a test for iron(III) ions, is the blood-red colour produced by the addition of thiocyanate ion, which may be discharged by the addition of fluoride ion with which the iron(III) ion complexes preferentially.

A curious feature of iron(III) is its inability to form stable complexes with nitrogen-donor ligands. Its affinity for oxygen-donor ligands, particularly if bidentate (e.g. acetylacetonate) is much greater. The coloured compounds formed with organic molecules containing hydroxyl groups are used widely in organic identification tests. The majority of its complexes are octahedral, but CFSE offers no extra stability for either high-spin octahedral or tetrahedral coordination and, as with Mn^{2+} (also d^5), a number of tetrahedral complexes such as $[FeCl_4]^-$ are known.

Ruthenium(III) is completely covalent in character and does not form cationic salts. The halides, with the exception of the fluoride, are known and in the case of the chloride gives rise to a wide range of halide complexes, $[RuCl_6]^{3-}$ and $[RuCl_5(H_2O)]^{2-}$ etc. Unlike iron(III), ruthenium(III) forms a large number of complexes with nitrogen-donor ligands. With ammonia, complex species are produced containing three, four, five, and six ammonia molecules. Osmium(III) also forms some halide and nitrogen-donor complexes, but these are few in number.

Oxidation state $+2$ (d^6)

This is one of the two most stable states for iron and is found with both the high-spin configuration and the diamagnetic t_{2g}^6 configuration. All ruthenium (II) and osmium(II) compounds are low-spin and many complexes of the former are surprisingly inert, no doubt because of the symmetrical t_{2g}^6 configuration. Stabilization of this state is also achieved for ruthenium and osmium by complexing with π-acceptor ligands such as carbon monoxide and nitric oxide.

Iron(II) hydroxide is quite strongly basic and gives rise to green, ionic salts with most anions. Fe^{2+} is mildly reducing and in this respect continues the sequence of diminishing reducing power found for first transition series M^{2+} ions (Mn^{2+} being anomalous because of its d^5 configuration). Aqueous solutions containing $[Fe(H_2O)_6]^{2+}$ are oxidized by air but much less rapidly in acid than in alkaline solutions from which the dark coloured tri-iron octahydroxide $Fe_3(OH)_8$ is precipitated. The extreme insolubility of this material is at least partly responsible for the influence of alkali on the ease of oxidation. A number of iron(II) complexes are known, but not so many as for iron(III). It is probably a fair generalization to say that where spin-pairing occurs (as with ligands such as CN^-), iron(II) (t_{2g}^6) is the more stable, but without spin-pairing, iron(III) ($t_{2g}^3 e_g^2$) is favoured. It is certainly noticeable that the hexacyanoferrate(II) $[Fe(CN)_6]^{4-}$ is more stable than the hexacyano-ferrate(III) $[Fe(CN)_6]^{3-}$. In the former, the cyanide ions are so tightly bound that the complex ion is non-poisonous, whereas in the latter, dissociation renders it quite toxic. With iron(III) ions, hexacyanoferrate(II) gives a deep-blue precipitate (Prussian Blue) identical with that produced by the action of iron(II) ions on hexacyanoferrate(III). It is formulated as $[Fe^{II}Fe^{III}(CN)_6]^-$ and its intense colour may be associated with the presence of iron in two oxidation states and the possibility of their interconversion by charge transfer. Treatment of hexacyanoferrate(II) or (III) with 30 per cent nitric acid gives dark-red crystals of the so-called *nitroprussides*, $[Fe(CN)_5NO]^{2-}$. Since these compounds are diamagnetic, the anion is most easily regarded as containing NO^+ coordinated to low-spin Fe^{2+}. However, the actual situation is probably more complicated, involving appreciable π-bonding. A number of related pentacyano complexes can be made, and the reactivity of the NO group is utilized in the test for sulphide ions which produce a reddish-violet colour due to $[Fe(CN)_5NOS]^{4-}$.

Although octahedral coordination is the most common for iron(II), several tetrahedral complexes are known. These are mostly halide complexes of the type $[FeX_4]^{2-}$ with a large cation.

Mention must be made of what is biologically one of the most important of all coordination compounds, haemoglobin **13**. This contains high-spin iron(II) coordinated to the four planar nitrogen atoms of a *porphyrin* ring

13
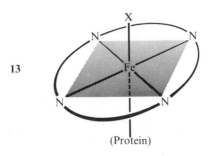
(Protein)

system, and attached also to a nitrogen atom of a protein. The sixth position (X) may be occupied by water, and the complex functions biologically because of the reversible replacement of this by molecular oxygen when the molecule becomes low-spin and diamagnetic. This is of course an oversimplification, and the way in which the oxygen is bonded to the iron is not yet understood. The ability of cyanide ions and carbon monoxide to replace the oxygen and so destroy the oxygen-carrying ability is the main reason for the extremely poisonous nature of these materials.

Lower oxidation states are stabilized in the carbonyls and related compounds. The noble-gas configuration is attained by five-coordination; the three pentacarbonyls $M(CO)_5$ are liquids, and their molecules have trigonal bipyramidal structures. Octahedral coordination can be achieved by polymerization, and the binuclear species $M_2(CO)_9$ are known for iron and osmium and the trinuclear $M_3(CO)_{12}$ for all three metals. $Fe_2(CO)_9$ **14** has three bridging CO groups and its diamagnetism implies the presence of an Fe–Fe bond.

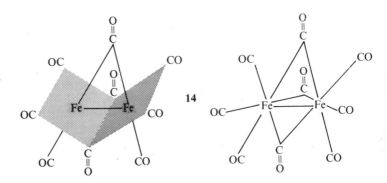

14

Among the multitude of organometallic compounds of iron is the first and best-known of the 'sandwich' compounds, ferrocene $Fe(C_5H_5)_2$ **15**. The molecule is most readily regarded as being composed of an Fe^{2+} ion attached

15

to two cyclopentadienide ($C_5H_5^-$) anions. If each carbon atom is considered to be sp^2 hybridized, forming bonds with the two adjacent carbons and one hydrogen, then a single electron in a p_z orbital perpendicular to the ring system remains. These p_z orbitals form five π molecular orbitals and the electrons they contain, plus the additional electron (because the ring is negatively charged) give three electron pairs occupying the three lowest molecular orbitals. Each ring may then be considered to be coordinated to the Fe^{2+} ion by donation of these three pairs of 'aromatic' π-electrons. The iron then achieves the noble-gas configuration and is diamagnetic.

PROBLEMS

3.8. Explain why aqueous solutions of manganese(II) salts are much less intensely coloured than those of many other transition-metal ions and why the same cannot be said of aqueous solutions of the iso-electronic iron(III) ion.

3.9. Aqueous iron(III) chloride and aqueous potassium chromate both separately liberate iodine slowly from potassium iodide solution, but when mixed together they produce a rapid liberation of iodine from the iodide solution. Why?

Cobalt, rhodium, and iridium   Co $3d^7 4s^2$, Rh $4d^8 5s^1$, Ir $5d^9$

Cobalt is the least plentiful element in the first transition series (< 0.004 per cent of the earth's crust) and is usually found associated with nickel. It is used in certain steels and also in ceramics to impart a blue colour. Rhodium and iridium are very rare but find limited uses as highly corrosion-resistant alloys.

Physical properties

	Co	Rh	Ir
Density/(g cm^{-3})	8·83	12·41	22·42
Melting power (°C)	1495	1966	2410
Boiling point (°C)	2900	∼3800	∼4500
Covalent radius/nm	0·116	0·125	0·126

In progressing across the transition series, the maximum oxidation state increases and reaches its highest value ($+8$) in the previous group. However, the effect of the increasing penetration of the d-orbitals into the inert electron core (Fig. 0.3, p. 5) is to stabilize them and so render more difficult the attainment of high oxidation states. Starting with the present group this effect begins to be dominant and the value of the maximum attainable oxidation state gradually decreases in the remaining groups. Oxidation states up to $+6$ are found for rhodium and iridium, but $+4$ is the highest found for cobalt, for which it is extremely unstable. The most common states are $+2$ and $+3$ for cobalt, $+3$ for rhodium and $+3$ and $+4$ for iridium. Cobalt is remarkable in that nowhere is the effect of complexing on the relative stabilities of oxidation states more marked. Complexes of cobalt(III) are amongst the most stable and numerous known, yet the simple Co^{3+} ion is reduced by water with evolution of oxygen. Conversely some complexes of cobalt(II) are oxidized by water with liberation of hydrogen.

Rhodium metal is rather soft but the other two are hard. They are comparatively inert, cobalt being less readily attacked by mineral acids than is iron. Heated in air they give different products: CoO and Co_3O_4, Rh_2O_3, and IrO_2. These are in fact the only oxides which can be obtained completely pure and their compositions reflect admirably the relative stabilities of the oxidation states of the three metals.

Oxidation state $+6$ (d^3)

The only compound of this state stable at room temperature is the highly reactive yellow solid IrF_6 which hydrolyses rapidly with atmospheric moisture. RhF_6 is even less stable. It is a black solid which loses fluorine at room temperature giving lower fluorides. Both these fluorides are octahedral.

Oxidation state $+5$ (d^4)

This state is confined to the fluorides of rhodium and iridium which, like the pentafluorides of ruthenium and osmium form tetramers, and complexes of the type $[MF_6]^-$. All are strongly oxidizing.

Oxidation state $+4$ (d^5)

For an element in the second half of the first transition series this oxidation state is too high to allow even a half-filled d-shell to stabilize it. As a result, the only examples of cobalt(IV) appear to be a few fluoro-complexes. Rhodium(IV) is little better, the fluoride and a few complexes of the type $M_2^I[RhF_6]$ being the chief examples. Iridium(IV) is considerably more stable and besides the oxide, complex ions $[IrX_6]^{2-}$ where X = F, Cl, and Br, and other hydrated species are found.

Oxidation state $+3$ (d^6)

This is the most prolific state for all three elements and is extremely stable providing the metal ion is complexed and has the low-spin t_{2g}^6 configuration.

But, as was noted above, Co^{3+} is unstable in water and only a few simple salts (e.g. fluoride and sulphate) can be isolated. Even these are unstable in aqueous solution unless a strong oxidizing agent is present. The importance of spin-pairing in stabilizing cobalt(III) can be seen in the fact that all its complexes are octahedral and its only high-spin compounds are salts of $[CoF_6]^{3-}$. The rest, even the unstable $[Co(H_2O)_6]^{3+}$, are diamagnetic and hence low-spin. Complexes with nitrogen-donor ligands are particularly stable though bidentate oxygen-donor ligands such as acetylacetonate and oxalate also complex strongly. The halides, which are low in the spectrochemical series and so do not favour spin-pairing, do not form such stable complexes whereas the hexacyanoanions are even more stable than those of chromium, manganese, or iron. With ammonia and amines a wide variety of polynuclear complexes of cobalt(III) are known which are held together by bridging groups such as OH^- or NH_2^-. The first compound not containing carbon ever to be resolved into its optically active isomers was one of these, **16**. Rather sur-

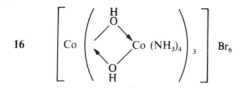

16 $\left[Co \left(\begin{array}{c} H \\ O \\ \diagup \diagdown \\ \diagdown \diagup \\ O \\ H \end{array} Co\ (NH_3)_4 \right)_3 \right] Br_6$

prisingly, a number of binuclear complexes, of which $[(NH_3)_5Co-O_2-Co(NH_3)_5]^{5+}$ is typical, are found to be paramagnetic with a magnetic moment of 1·7 B.m., suggesting the presence of a single unpaired electron. It was originally thought that one cobalt was $+3$ and the other $+4$. However, the phenomenon occurs only with $-O_2-$ bridges, and it seems preferable to regard the compounds as derivatives of cobalt(III) with the unpaired electron situated on the bridging group.

As with chromium(III), the importance of cobalt(III) in early work on co-ordination chemistry arose not only because of the large number of complexes produced but also from their kinetic inertness, which allows the separation of isomeric forms where these occur. This inertness can also make interconversion of cobalt(III) complexes difficult and so they are commonly made by oxidation of the more labile cobalt(II) in the presence of appropriate ligands. Very mild oxidizing agents, such as a stream of air bubbled through the solution, are often sufficient to effect this.

A variety of simple compounds of rhodium(III) and iridium(III) is known, such as the halides and sulphates. Both metals form a large number of complexes with ammonia, amines, and cyanides and, unlike cobalt, complex halides which are stable and diamagnetic. Indeed, as would be expected with 4d and 5d configurations, the tendency to spin-pairing is even more marked than with cobalt(III) and no high-spin compounds are known.

Oxidation state +2   (d^7)

The contrast between cobalt and the two heavier elements is very marked. Simple salts of cobalt(II) with all the common anions are stable, whereas rhodium(II) and iridium(II) form few simple compounds and usually require stabilization with π-acceptor ligands such as phosphines and arsines.

In the presence of weak coordinating agents such as water, halides, and

$$Co \xrightarrow{-0.277V} Co^{2+} \xrightarrow{1.82V} Co^{3+}$$

other common anions, cobalt(II) is stable against oxidation to cobalt(III). However, in basic solutions or when stronger coordinating agents are present, oxidation is easy. Complexes of cobalt(II), though well-known, are nothing like so numerous as those of cobalt(III) and this may be ascribed to the fact that those ligands which give strong M–L bonds facilitate oxidation to cobalt(III), while those ligands which favour the lower oxidation state do so because they give weaker M–L bonds. It is not surprising that low-spin octahedral complexes such as $[Co(CN)_6]^{4-}$ are unstable to oxidation.

Many cobalt(II) salts exist in two forms, one pink, one blue. The former are generally favoured by high dilution in aqueous solution and by extreme hydration in the crystalline form. The difference is due to a change in the coordination number of the metal, the pink colour arising from the octahedral ion $[Co(H_2O)_6]^{2+}$ and the blue from tetrahedral species such as $[CoCl_4]^{2-}$. The ease of formation of tetrahedral complexes with a d^7 cation is a direct consequence of the CFSE values mentioned in Chapter 2. A variety of anionic tetrahedral complexes of the type $[CoX_4]^{2-}$ as well as neutral complexes CoL_2X_2 is known, where X is a halide or SCN^- and L is a neutral ligand. The distinction between pink octahedral and blue tetrahedral complexes is a common, though not a universal one. A further distinction is provided by the magnetic moments. In high-spin configurations, both octahedral and tetrahedral stereochemistries lead to three unpaired electrons. For reasons which need not be elaborated here, both have magnetic moments higher than the spin-only value but those of the octahedral complexes are rather higher (4.8–5.2 B.m.) than those of the tetrahedral (4.4–4.8 B.m.).

There are a number of four-coordinate cobalt(II) complexes with magnetic moments somewhat higher than the spin-only value for one unpaired electron. These are presumably square-planar and include the dimethylglyoximate as well as complexes with some porphyrin-type ligands which exhibit oxygen-carrying ability not unlike that found for iron. Vitamin B_{12} is a related complex of cobalt whose redox properties are apparently utilized in the biological synthesis of the ligands attached to iron in haemoglobin.

Lower oxidation states are found in complexes with π-acceptor ligands. The M(I), d^8 complexes are most often five-coordinate and low-spin (therefore diamagnetic) and the carbonyls of the M(0), d^9 metals are of necessity binuclear if the noble-gas configuration is to be attained. $Co_2(CO)_8$ exists in two isomeric forms (**17**). In **17a** the molecule is 'folded' in order to allow overlap of the metal orbitals, while **17b** is similar to $Mn_2(CO)_{10}$ in having no bridging groups. The position with rhodium and iridium carbonyls is less clear, but all three metals form higher polymers such as $M_4(CO)_{12}$.

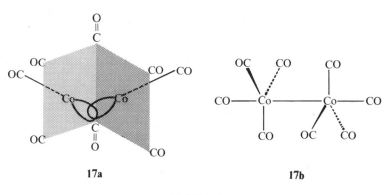

17a   **17b**

PROBLEM

3.10. Cobalt(II) forms both octahedral and tetrahedral complexes. Electrolysis of aqueous solutions where both occur gives a pink colour at the cathode and a blue colour at the anode. Furthermore the spectrum of the cathodic species is independent of the anion, whereas that of the anodic species varies with the anion. Explain.

Nickel, palladium, and platinum   Ni $3d^8 4s^2$, Pd $4d^{10}$, Pt $5d^9 6s^1$

Nickel is not particularly common, being about twice as abundant as cobalt and constituting about 0·1 per cent of the earth's crust. It is usually combined with sulphur, arsenic, or antimony. Its extraction is usually effected by separation of the sulphide, which is heated in air to give the oxide, which is then reduced with carbon to give the metal. The original Mond process is still used to some extent, particularly to produce high-purity nickel, and is interesting because it depends on use of the volatile carbonyl. The ores after conversion to the oxides are reduced with water gas (hydrogen and carbon monoxide) at a temperature of 300°C which is too low to reduce other contaminating oxides. Treatment with carbon monoxide at 50°C gives nickel carbonyl $Ni(CO)_4$ which is then decomposed at about 200°C. Nickel is used in the very tough nickel steel, as a protective coating of iron, and in many alloys especially for coinage purposes. Palladium and platinum are rather rare but, along with nickel, are very efficient catalysts especially for hydrogenation and dehydrogenation reactions.

Physical properties

	Ni	Pd	Pt
Density/(g cm^{-3})	8·90	12·02	21·45
Melting point (°C)	1453	1552	1769
Boiling point (°C)	2732	2927	~3800
Covalent radius/nm	0·115	0·128	0·129
Crystal radius/nm			
M^{2+}	0·073	0·085	0·096
M^{3+}	...	0·074	0·083
M^{2+}, M $E^{\ominus}$/V (see p. 96)	−0·25	+0·987	~ +1·2

The unwillingness of transition metals at the end of a series to exhibit high oxidation states and the increasing stability of the lower oxidation states, particularly of the heavier elements, is again evident in this group. Only in the penta- and hexafluorides of platinum are oxidation states higher than +4 found. The most common are +2 for nickel; +2 and to a lesser extent +4 for palladium; +2 and +4 for platinum. Only nickel(II) shows an extensive cationic chemistry, although aquated Pd^{2+} occurs with non-coordinating ions such as perchlorate. Platinum completes the trio of elements most widely investigated by early coordination chemists, its +2 and +4 compounds being comparable with those of chromium(III) and cobalt(III) in stability and kinetic inertness.

The pure metals are comparatively unreactive, though palladium and platinum are somewhat more reactive than the other platinum metals. They are resistant to atmospheric corrosion but in the case of nickel this seems to be due to a protective oxide film; when very finely divided it is pyrophoric. It reacts only slowly with dilute hydrochloric and sulphuric acids, and like iron becomes passive if treated with concentrated nitric acid. Palladium is attacked by hot concentrated nitric acid but platinum only by aqua regia. Their resistance to alkaline attack is in the reverse order, palladium and platinum being attacked rapidly by fused alkali but nickel only very slowly. It is for this reason that nickel rather than platinum vessels are used in the laboratory for alkali fusions. Nickel or its alloys are also used in the construction of apparatus for performing fluorinating reactions because, although all three metals react with fluorine and chlorine when heated, the nickel–fluorine reaction is only slow.

The metals show a remarkable power to absorb hydrogen gas. This is most marked for palladium which is used as a diffusion membrane for purifying hydrogen. Though the reasons for the phenomenon are not fully understood, it is clear that it is at least partly responsible for the efficiency of these elements as hydrogenation catalysts.

The only well-characterized oxides are NiO, PdO, and PtO_2, of which only PdO is conveniently prepared from the elements. Green nickel(II) oxide is obtained by heating nickel(II) hydroxide, carbonate, or nitrate, and the black platinum(IV) oxide by gently heating (strong heating drives off oxygen), or otherwise dehydrating, the hydrate precipitated by the addition of alkali to platinum(IV) solutions. Addition of alkali to palladium(IV) and platinum(II) solutions precipitates hydrated PdO_2 and PtO respectively. The former is strongly oxidizing, and the latter strongly reducing, and neither can be dehydrated without decomposition.

Oxidation states $+6$ *and* $+5$ $(d^4$ *and* $d^5)$

PtF_6, PtF_5, and possibly some oxy-fluorides are the only examples of these high oxidation states. They are extremely reactive materials, hydrolysed rapidly by moisture.

Oxidation state $+4$ (d^6)

All the compounds in this oxidation state are uniformly octahedral and diamagnetic, and hence low-spin. A few complexes of nickel(IV) are known including a number of mixed oxides, and the complex K_2NiF_6. Compounds of palladium(IV) are confined to the hydrated oxide, the sulphide and a few complex halides such as $[PdCl_6]^{2-}$. In contrast, platinum(IV) forms many compounds and is virtually as stable as platinum(II). The amine and halide complexes are particularly numerous, covering the range from $[Pt(am)_6]^{4+}$ and $[Pt(am)_5X]^{3+}$ to $[PtX_6]^{2-}$, and being particularly stable when X = Cl. The series with ammonia and chloride was used by Werner to establish the coordination number of platinum(IV) as six. The conductivities are roughly proportional to the number of ions and most significantly the conductivity is zero for $Pt(NH_3)_2Cl_4$.

Dissolution of metallic platinum in aqua regia yields the acid H_2PtCl_6. Addition of alkali forms a series of complexes as the chloride ions are replaced stepwise by hydroxide, culminating in the 'hydrated oxide' $H_2[Pt(OH)_6]$.

Oxidation state $+3$ (d^7)

Only nickel is known with certainty to show this oxidation state and even then the compounds are few. Though no anhydrous oxide of nickel other than NiO can be obtained, hypochlorite (chlorate(I)) oxidation of alkaline nickel(II) solutions gives a black unstable precipitate of variable composition which contains nickel in an oxidation state higher than two. More satisfactory evidence for nickel(III) is provided by the existence of K_3NiF_6 and some complexes with polarizable arsenic-donor ligands.

Certain compounds once thought to contain palladium(III) are now known to contain palladium(II) and (IV). An example of this is the black crystalline 'trifluoride' which is actually $Pd^{2+}[PdF_6]^{2-}$.

Oxidation state +2   (d^8)

This is the only oxidation state stable for all three elements. Nickel(II) forms simple salts with all the common anions, aqueous solutions of which contain the green hydrated ion $[Ni(H_2O)_6]^{2+}$. As is to be expected, palladium(II) has a more limited cationic chemistry and platinum(II) is always complexed. The stability of the complexes increases down the group, and the two heavier elements are notable for the propensity with which they form square-planar compounds. In these (see Chapter 2) the destabilization of the d_{z^2} orbital is so great that the eight d-electrons are forced to pair in the remaining four d-orbitals, making the complexes diamagnetic. Nickel(II) also forms square-planar complexes but has, to a unique extent, the ability to adopt a variety of stereochemistries. The facility with which interconversion between these is possible has led to considerable confusion.

As well as in the aquo-ion, octahedral coordination occurs for nickel(II) with ammonia and amines, the Ni—N bond being much stronger than Ni—O. Tetrahedral complexes are formed mainly with halide ions, i.e. $[NiX_4]^{2-}$, in which the stereochemistry is stabilized by large cations such as $[Ph_4P]^+$, and with certain phosphines and arsines. The most stable four-coordinate complexes however are square-planar. Typical of these are $[Ni(CN)_4]^{2-}$ and the insoluble complex **18** with dimethylglyoxime which is

18

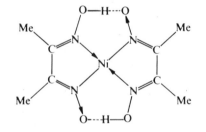

used in the qualitative and quantitative estimation of nickel. In general, the octahedral and tetrahedral complexes are green or blue and paramagnetic, while the square-planar are red or yellow and diamagnetic. However, a number of complexes exist, particularly with ligands containing both oxygen and nitrogen donor atoms, which have magnetic moments greater than zero but well below those expected for two unpaired electrons. In solution the moments frequently depend on solvent and concentration, and it appears that the interconversions:

$$\text{square-planar} \rightleftharpoons \text{tetrahedral}$$

$$\text{square-planar} \underset{}{\overset{\text{solvent coordination}}{\rightleftharpoons}} \text{octahedral}$$

are taking place. Indeed, many compounds formally described as square planar or tetrahedral are in fact intermediate between the two. This is easily

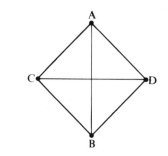

19

understood from **19** which is a view of a tetrahedron, the apices of which (A, B, C, and D) may be taken to represent four ligands. Rotation of A–B through 90° produces a square-planar arrangement while any intermediate rotation produces a configuration which only approximates to one or other of the extremes.

An interesting compound prepared by the action of ammonia and benzene on nickel(II) cyanide is $Ni(CN)_2 \cdot NH_3 \cdot xC_6H_6$ where $x < 1$. This is a *clathrate* compound consisting of a rather complicated polymeric lattice, inside which the benzene molecule is trapped as in a cage. The coordination of half the nickel ions is octahedral and of the other half square-planar, giving a magnetic moment equivalent to an average of one unpaired electron per nickel.

The vast majority of palladium(II) and platinum(II) compounds are four-coordinate, square-planar, and diamagnetic although a number of their reactions are believed to proceed via five-coordinate species. Coordination is weak to oxygen and fluorine but very strong to nitrogen and cyanide. It is notable that for compounds of the type MX_2Y_2 only one form is possible if the coordination is tetrahedral whereas two, known as *cis* and *trans* isomers, are possible if the coordination is square-planar (**20**).

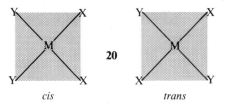

cis   **20**   *trans*

Isolation of two isomers of the non-electrolyte $[Pt(NH_3)_2Cl_2]$ by Werner led him to the belief that platinum was square-planar. Although now known to be correct, this assignment was not unequivocal because of the possibility of an isomeric five-coordinate dimer $(NH_3)_2ClPt$
$$\begin{array}{c} Cl \\ \diagdown \diagup \\ \diagup \diagdown \\ Cl \end{array}$$
$PtCl(NH_3)_2$. A further

isomer, Magnus's green salt, was already known. This is the electrolyte $[Pt(NH_3)_4]^{2+}[PtCl_4]^{2-}$ in which both ions are square-planar.

It is usual to make 'horizontal' comparisons between the coordination chemistry of nickel(II) and that of other M(II) ions of the first transition series. However, the existence of ostensibly similar complexes of all the M(II) ions of this group makes 'vertical' comparisons inevitable. The most obvious difference is that whereas the crystal-field approach provides a reasonable basis for describing the bonding and accounting for the spectra of many of the complexes of nickel(II) and other first-series M(II) ions, it does not do so for the complexes of palladium(II) and platinum(II). In these cases the bonding is too covalent, even in such complexes as $[MCl_4]^{2-}$, and any serious attempt at explanation must invoke molecular-orbital theory.

Of the lower oxidation states, only zero is of any importance. π-acceptor ligands are required for its stabilization, but it is significant that the tetrahedral $Ni(CO)_4$ is the sole binary carbonyl. It appears that for palladium and platinum the $(n-1)$d electrons have sufficiently penetrated the inert electron core to be capable of only limited involvement in back-donation to the ligands. One or two CO groups can be attached to these atoms but they must be accompanied by other ligands, such as phosphines, with more pronounced σ-donor capability.

Copper, silver, and gold $(n-1)d^{10}ns^1$

Copper is rather less common than nickel; the other two elements are rare. All three occur native and copper and silver often as sulphide ores. Copper is usually recovered by aerial oxidation of these ores, the mixture of Cu_2O and Cu_2S then being heated to drive off sulphur dioxide, leaving the impure metal. For silver and gold, treatment with sodium cyanide solution and aeration produces the cyano-complexes $[M(CN)_2]^-$ from which the metals are precipitated by adding zinc to the alkaline solutions. Electrolytic purification is possible for each of the metals. They are collectively known as the 'coinage' metals, though this description is now outdated. Copper is widely used as an electrical conductor and also in alloys (with zinc as brass and with tin as bronze). Because of their cost, silver and gold are not so widely used as their properties would allow.

Physical properties

	Cu	Ag	Au
Density/(g cm^{-3})	8·96	10·50	19·32
Melting point (°C)	1083	961	1063
Boiling point (°C)	2595	2212	2966
Covalent radius/nm (tetrahedral)	0·135	0·153	0·150
M$^+$ crystal radius/nm	0·093	0·121	0·137

The relevant electrode potentials are summarized below:

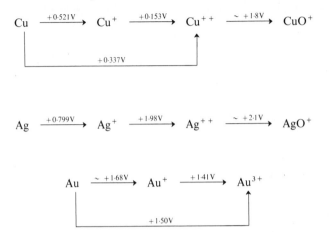

The pure metals are fairly soft, extremely malleable and ductile, and excellent conductors of heat and electricity. Gold is so malleable that ordinary gold leaf has a thickness of only 10^{-7} m (100 nm or about 350 atoms). Chemically they are unreactive. Being below hydrogen in the electrochemical series they do not react with non-oxidizing acids in the absence of air, and indeed gold is attacked only by aqua regia. Copper reacts with oxygen, copper and silver with sulphur, and all three with cyanides if air is present.

However, this is in many ways the most perplexing group of the transition elements. Whilst the pure metals are physically quite similar, their chemistries are not and the differences are difficult to systematize. For each element one oxidation state is much more common than any other. For copper this is $+2$; for silver $+1$; for gold $+3$. That $+3$ is the highest known oxidation state, and is stable only for the heaviest element, is understandable for a group at the end of the transition series, but the erratic sequence of the stable oxidation states is more difficult to explain. The sequence is in accord with the observed ionization energies. The first ionization energy is smallest for silver; the sum of the first and second is smallest for copper; the sum of the first, second and third

Ionization energies/(kJ mol^{-1})

	1st	2nd	3rd
Copper	747	1957	3554
Silver	731	2073	3360
Gold	891	1979	2941

is smallest for gold. Unfortunately, this is not so much an explanation as a restatement of the problem, and it is clear that the energies of the ns and $(n-1)d$ orbitals and also the effect of ionic charge on these, vary in a complicated way between the three elements.

Oxidation state $+3$ (d^8)

As with the d^8, M^{II} species of the previous group, there is a marked tendency to form diamagnetic, square-planar complexes, $[CuF_6]^{3-}$ being the only one known to be paramagnetic and octahedral. Only for gold is this state common; not as the simple Au^{3+} cation (the halides MX_3 are polymerized and the metal has a coordination number of four) but coordinated to a variety of donor atoms. These include the halides, carbon, arsenic, and, in contrast to Pd^{II} and Pt^{II}, oxygen. Dissolution of gold in aqua regia gives the acid $HAuCl_4$ from which a series of salts may be prepared. Treatment with alkali precipitates the hydroxide, $Au(OH)_3$, which on gentle heating gives the brown Au_2O_3. A general feature of the chemistry of gold(III) is the relative ease with which reduction to the metal (often in a highly coloured colloidal form) may be produced with quite mild reducing agents.

Oxidation state $+2$ (d^9)

This is virtually unknown for gold, strongly oxidizing for silver, but is the best known oxidation state for copper. Compounds purporting to contain gold(II) are invariably mixtures of gold(I) and gold(III). This phenomenon is also found in the black, diamagnetic oxide, formed by the oxidation of silver(I) solutions with alkaline persulphate: it is in fact $Ag^IAg^{III}O_2$ rather than $Ag^{II}O$. The simple Ag^{2+} cation oxidizes water and must be complexed to be rendered stable. This is possible with a number of nitrogen-donor ligands such as pyridine and bipyridyl which produce four-coordinate, square-planar complexes. Silver(II) fluoride which is readily hydrolysed by moisture is the only binary compound of silver(II).

Black copper(II) oxide, obtained when copper is heated in air, and the blue hydroxide precipitated when alkali is added to aqueous solutions of copper(II), are mainly basic and lead to a series of salts of the cation Cu^{2+}. These salts are either blue or green and are paramagnetic with magnetic moments corresponding to one unpaired electron. In aqueous solutions they produce the blue aquo-ion $[Cu(H_2O)_6]^{2+}$, which has a distorted octahedral structure with two *trans* water molecules less strongly bonded than the remaining four planar water molecules. It will be recalled from Chapter 2 that the d^9 configuration is expected to favour square-planar or distorted octahedral structures. Copper(II) hydroxide is readily soluble in aqueous ammonia giving the strongly basic $[Cu(NH_3)_4(H_2O)_2]^{2+}(OH^-)_2$ with a characteristic intense blue colour.

The anhydrous halides of copper(II) afford an interesting series. Fluorine, which is very electronegative, gives rise to ionic CuF_2 with a distorted rutile structure. The less electronegative chlorine and bromine, being unable to effect complete electron transfer, give rise to covalent chain structures containing square-planar CuX_4 units, **21**. Finally, iodine is unable to involve in

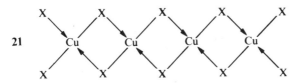

21

bonding more than one electron of the copper, and only copper(I) iodide is stable; iodide ions reduce Cu^{2+} to give copper(I) iodide and free iodine, the reaction being assisted by the insolubility of copper(I) iodide. In a similar way, cyanide ions yield copper(I) cyanide and cyanogen.

Copper(II) nitrate, if prepared by the usual aqueous methods, crystallizes as the trihydrate. The anhydrous salt can be prepared by reacting copper with a solution of dinitrogen tetroxide in ethyl acetate to give $Cu(NO_3)_2 \cdot N_2O_4$ which loses the N_2O_4 at 85°C. It is remarkable in that it can be sublimed *in vacuo* and was the first example of a metal nitrate to do so. It is monomeric in the vapour and the nitrate must therefore be covalently bonded. The precise nature of the bonding is still not certain and for this reason has been omitted in **22**, which merely shows the spatial distribution of the molecule in a simplified manner:

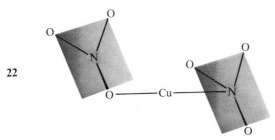

22

Copper(II) coordinates strongly to nitrogen- and oxygen- donor ligands and to halides. As indicated by the Irving–Williams order, coordination is generally stronger for copper(II) than for any other divalent first series ion, but, because of the Jahn–Teller effect, this applies only to the first four ligands and not to the fifth and sixth. The crystal-field approach provides a good description of d^9 complexes and, as was noted for the copper(II) halides, this configuration becomes less stable as the bonding becomes more covalent. Tetracyanocuprate(II), $[Cu(CN)_4]^{2-}$, is thus unknown.

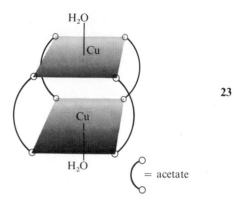

23

A feature of copper(II) chemistry is the formation of binuclear molecules. The best known of these is the acetate **23** in which the binuclear structure is maintained by means of acetate bridges. A magnetic moment less than that expected for one unpaired electron per copper, indicates the presence of weak metal–metal bonding, i.e. the electron spins are partially paired. Note the close similarity with chromium(II) acetate (p. 63).

In spite of the prevalence of square-planar and distorted octahedral stereo-chemistries, the possibility of other arrangements cannot be dismissed. The addition of excess halide ions to solutions of copper(II) halides causes the colour to change from blue towards yellow. Salts of the anions $[CuX_4]^{2-}$ may be isolated and, depending on the associated cations, are found to involve coordination around the copper varying from almost planar to almost tetrahedral (compare nickel, p. 84).

Oxidation state +1   (d^{10})

In the original Periodic Table of Mendeleev, the alkali and coinage metals were subgroups A and B respectively of Group I. For this reason alone it was obligatory to look for comparisons, and the existence (but not the chemistry) of this oxidation state provided the main justification for the classification. The presence of the completely filled d-shell implies some similarity with the noble-gas configuration of the alkali-metal ions. Certainly the compounds of the d^{10} ions are uniformly diamagnetic and, unless the anion is itself coloured or encourages charge transfer, are colourless. However, in the neutral atoms the screening of the ns^1 electron from the attraction of the nucleus is performed less effectively by the d-shell, and the formation of M^+ is more difficult for the coinage metals. On the other hand, the filled d-shell is more easily ruptured so that further ionization is much easier. This can be seen by comparing the first and second ionization energies of sodium (496 and 4564 kJ mol^{-1}) with those of copper silver and gold given earlier. So far as their chemistries are concerned, the d^{10} ions form complexes far more readily and their compounds show a greater degree of covalency than do those of the alkali metals.

This oxidation state is much more stable for silver than for either copper or gold. The apparent instability of copper in aqueous solution is at least partly explained by the high heat of hydration of copper(II) and the low solubility of copper metal, which favour the disproportionation

$$2Cu^I = Cu^{II} + Cu^O \downarrow \quad (K = 1\cdot2 \times 10^6 \text{ at } 25°C).$$

In practice, copper(I) is stable in very insoluble salts or if appropriately complexed. The insoluble compounds include the oxide Cu_2O (more stable at high temperatures than CuO), sulphide, halides, and cyanide, in all of which appreciable covalency is involved. In complexes, the stereochemistry is often tetrahedral, and $[Cu(CN)_4]^{3-}$ is more stable than complexes with only two or three coordinated cyanides.

Silver(I) compounds are numerous, but there is little tendency to coordinate to oxygen and the salts are usually anhydrous. With the exception of the fluoride, nitrate, and perchlorate they are also insoluble in water and are often appreciably covalent. The colours of the halides vary from white (chloride) to yellow (iodide) the shift of the charge-transfer absorption into the lower-energy, visible region of the spectrum reflecting the increasing covalency. The fluoride and perchlorate are remarkable for their great solubility in water and, in the case of the latter, in non-polar solvents (even hydrocarbons) also. As with copper(I), coordination numbers up to four are possible, but for silver(I) a coordination number of two is much more common, and particularly stable compounds are formed with cyanide and with sulphur-donor ligands. Of some use in qualitative analysis is the fact that while silver(I) chloride may be dissolved in aqueous ammonia solution to give $[Ag(NH_3)_2]^+$ ($\beta = 1\cdot6 \times 10^7$), the less soluble silver(I) iodide. requires cyanide or thiosulphate for dissolution, giving the much more stable $[Ag(CN)_2]^-$ ($\beta = 5\cdot6 \times 10^{18}$) and $[Ag(S_2O_3)_2]^{3-}$ ($\beta = 1\cdot7 \times 10^{13}$). (The stability constant β was defined on p. 32.)

Gold(I) is rather similar to copper(I) except that it is stable only when complexed and that disproportionation is to the metal and gold(III) rather than to gold(II). As with silver(I), a coordination number of two is favoured.

The chemistry of these elements does not appear to extend below the $+1$ oxidation state and in particular there are no simple carbonyls. This is not surprising since the same is true of palladium and platinum and in this group the d-electrons are even less likely to participate in extensive back donation to the ligands. Nevertheless, ammoniacal solutions of copper(I) chloride will absorb carbon monoxide quantitatively and $CuCl\cdot CO\cdot H_2O$, a colourless diamagnetic compound, can be isolated. It is a dimer with two chloride bridges. The monomer $AuCl\cdot CO$ can also be obtained from non-aqueous media, but there is no silver analogue. A number of similar types of complex is found with unsaturated hydrocarbons such as ethylene.

Zinc, cadmium, and mercury $(n-1)d^{10}ns^2$

Constituting about 0·02 per cent of the earth's crust, zinc may be considered moderately abundant, but cadmium and mercury are rare. However, all three elements are familiar because of the ease with which they are extracted from their ores. The common ores are sulphides and roasting converts these to the oxides from which zinc and cadmium are obtained by reduction with coke. In the case of mercury(II) oxide, the metal distils without the need for a reducing agent. Zinc and cadmium are widely used as protective coatings and as alloys, of which brass (zinc and copper) is the best known.

Physical properties

	Zn	Cd	Hg
Density (g cm^{-3})	7·13	8·65	13·55
Melting point (°C)	419	321	-39
Boiling point (°C)	907	765	357
Covalent radius/nm (tetrahedral)	0·131	0·148	0·148
M^{2+} crystal radius/nm	0·072	0·096	0·110
M^{2+}(aq), M $E^{\ominus}$/V (see p. 96)	$-0·763$	$-0·403$	$+0·854$

The pure metals are notable for their low melting and boiling points and for their volatility. Mercury is the only metal which is liquid at room temperature and its vapour can present a serious health hazard. Many metals dissolve in mercury to form *amalgams*. This mutual solubility is common among metals above their melting points, and is important here because of mercury's low melting point. These amalgams are often useful as reducing agents and some of them must involve chemical combination. Hg_2Na and Hg_2K for example have definite compositions, high heats of formation, and melting points of 346 and 270°C respectively—much higher than those of their components.

Zinc and cadmium are higher in the electrochemical series than hydrogen, and so dissolve in dilute acids with liberation of hydrogen. Mercury is lower, and dissolves only in oxidizing acids such as nitric and concentrated sulphuric. Zinc alone reacts with boiling aqueous alkali. All three metals react with halogens, oxygen, and sulphur but none with hydrogen, carbon, or nitrogen.

In this group the $(n-1)d$ shells are full and are firmly entrenched in the inert electron core. The $(n-1)d$ electrons do not directly participate in bonding, $+2$ being the highest oxidation state, and the elements are not strictly transition metals at all. In Mendeleev's Periodic Table they were designated as Group IIB and comparison with Group IIA (the alkaline-earth metals) is more profitable than between Groups IA and IB. In particular, there are many similarities between magnesium on the one hand and zinc and cadmium on the other, and the last two elements are more electropositive than the preceding elements at the end of the transition series. With the exception of mercury, the chemistry of the group is virtually confined to the group oxidation state of two, corresponding to the loss or involvement in covalent bonding of the ns^2 electron pair, and back donation to ligands is even less likely to occur than with the coinage metals. Consequently carbonyls and alkene complexes do not occur. All their compounds, even those of mercury(I), are diamagnetic because they all contain a filled $(n-1)d$ shell. However, the presence of this filled d-shell cannot be ignored. It is more easily distorted than the noble-gas configuration of Group IIA M^{2+} ions and, as is characteristic of transition-metal ions in low oxidation states (p. 10), covalency increases down the group even though the size of the ions increases. Indeed covalency is more character-istic of the IIB elements generally since their s-electrons are less effectively shielded from the nuclear charge by the d^{10} shell than are the s-electrons of the IIA elements by the noble-gas shell. It may therefore be considered that the IIB atoms are less easily ionized, or alternatively that their M^{2+} ions are smaller than those of the IIA elements, and so form more covalent bonds.

In the case of mercury the even poorer shielding by the 4f electrons en-hances still more the stabilization of the $6s^2$ electrons, leading to the so-called *inert-pair* effect, important in the succeeding elements thallium and lead. This accounts at least in part for the differences between mercury and the other two elements in its group. The simplest manifestation of the inert-pair effect for mercury is perhaps the fact that it is a liquid at room temperature: the atoms are difficult to oxidize and are also unwilling to contribute electrons for metallic bonding (see p. 17).

All three oxides, MO, are formed by direct combination but differ in their stabilities. Zinc oxide and cadmium oxide can be sublimed without decompo-sition whereas mercury(II) oxide dissociates above about 400°C. The oxide and hydroxide of zinc are amphoteric and with alkali readily give zincates such as $Na_2[Zn(OH)_4]$. Corresponding cadmiates are produced only with difficulty using very concentrated alkali, and mercury(II) oxide (the hydroxide is unknown), though only weakly basic, is more basic than acidic. Fluorides, nitrates, sulphates, and perchlorates, are ionic, but the cations particularly mercury(II), are appreciably hydrolysed in water. The other halides are covalent, as is shown by their lower melting points and their solubilities in organic solvents, and those of mercury exist in aqueous solution almost entirely as undissociated HgX_2 molecules.

Metal–carbon bonds are found in the classes of compounds represented by RMX (R = alkyl or aryl, X = halide, M = Zn or Hg) and R_2M (M = Zn, Cd, or Hg). The RZnX compounds are analogous to the Grignard reagents of magnesium, but less reactive because of the lower affinity of zinc for oxygen, and RHgX is not even affected by water. R_2M are non-polar, and hence linear molecules. Those of zinc react vigorously with oxygen and are often spontaneously inflammable; those of cadmium are rather less reactive, and those of mercury are stable to both air and water. This lack of reactivity of the Hg–C bond arises not from its inherent stability, indeed the bond is quite weak and is thermally unstable, but from the weakness of the Hg–O bond.

All three elements coordinate to halides, cyanides, and nitrogen-donor ligands, as do the transition elements, but it is notable that zinc, like magnesium but less markedly, shows a preference for fluorine and oxygen whereas cadmium coordinates more strongly to chlorine and sulphur, and mercury to nitrogen and sulphur. This trend is undoubtedly connected with the increasing tendency to covalency in passing down the group, and mercury shows a marked reluctance to coordinate to fluorine and oxygen, whose high electronegativities favour more ionic bonding. Whereas water-soluble salts of zinc and cadmium are usually hydrated, the hydrated salts of mercury are almost entirely confined to its salts with oxy-anions, where the water is associated with the anion.

Having the d^{10} configuration, the M^{2+} ions afford no crystal-field stabilization energy and the observed coordination numbers depend on the size and polarizing power of the cations, while the stereochemistries are those which minimize the repulsions between the ligands. The ionic radius increases down the group and, if only electrostatic forces were involved, it would be expected that the ability to accommodate more ligands would parallel this. Accordingly, Cd^{2+} attains octahedral six-coordination more readily than Zn^{2+}, but the trend does not continue to Hg^{2+} which rarely exceeds a coordination number of four. The reason for this is the increased tendency of Hg^{2+} to form covalent bonds, as a result of which linear two-coordination and tetrahedral four-coordination are the most common.

Halide complexes of the types $[MX_3]^-$ and $[MX_4]^{2-}$ are known for each metal. Fluorides occur only for zinc and the most stable of all is $[HgI_4]^{2-}$ formed from the most polarizable anion and the most strongly polarizing cation. Similar and more stable cyanide complexes occur. With ammonia, cationic species $[M(NH_3)_x]^{2+}$ are produced of which the commonest have $x = 4$, though $x = 6$ for zinc and cadmium, and $x = 2$ for mercury are also known. In the case of mercury a competitive reaction takes place involving the replacement of hydrogen attached to the nitrogen to produce, for instance, $Cl—Hg—NH_2$. If mercury(II) oxide in its finely divided yellow form is treated with aqueous ammonia the yellow Millon's base, $[NHg_2]^+OH^-·2H_2O$ results. The cation has a three-dimensional polymeric structure in which each mercury atom is linearly bonded to two nitrogen atoms and each nitrogen is tetrahedrally bonded to four mercury atoms. Salts are readily formed by the replacement of the hydroxide ion with other anions. Nessler's reagent is an alkaline solution of potassium tetraiodomercurate(II) K_2HgI_4, used as a sensitive test for ammonia with which it gives a brown coloration or precipitate due to the formation of the iodide of Millon's base. It is a measure of the strength of the Hg–N bond that this reaction occurs in spite of the stability of the $[HgI_4]^{2-}$ ion.

Only for mercury is a second oxidation state, apparently of $+1$, well established, and even then, since the salts all contain the $[Hg-Hg]^{2+}$ ion, the oxidation state is really $+2$, each mercury atom being covalently bonded to another. The equilibrium $Hg_2^{2+} \rightleftharpoons Hg^{2+} + Hg$ is rather evenly balanced and the oxidation and reduction are easily accomplished. Mercury(I) salts are decomposed by any reagent which disturbs the equilibrium, e.g. by precipitating insoluble mercury(II) compounds. Mercury(I) cyanide, hydroxide, and sulphide are therefore unknown. Apart from the nitrate, chlorate, and perchlorate, which are soluble, and the fluoride which is decomposed by water, the salts of mercury(I) are insoluble in water. The usual test for mercury(I) is treatment with aqueous ammonia which produces a black material. This consists of metallic mercury and mercury(II) compounds such as $Cl-Hg-NH_2$ mentioned above.

PROBLEMS

3.11. The vapour of mercury(I) chloride has a density corresponding to the formula HgCl, and spectral evidence indicates the absence of any Hg–Hg bonds. The vapour is however diamagnetic. Explain.

3.12.

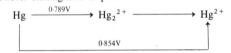

From the electrode potentials given in the scheme shown above, calculate the equilibrium constant for the reaction

$$Hg + Hg^{2+} \rightleftharpoons Hg_2^{2+}$$

Appendix: electrode potentials

WHEN a metal dips into a solution containing the cation of the metal, two processes take place simultaneously. Cations leave the solution and deposit on the metal, giving it a positive charge; and metal atoms leave the metal as cations and go into solution, giving the metal a negative charge. These processes initially take place at different rates, so that a potential difference is established between the metal and the solution. This potential difference is called the electrode potential of the metal. Conventionally we give the potential relative to the metal, so that the more readily the metal forms solvated cations, the more negative is its electrode potential. Only potential *differences* can be measured, and the standard hydrogen electrode is defined as having zero potential at all temperatures, other potentials being measured relative to this arbitrary zero. Values of E quoted in this book refer to acid solutions.

The standard electrode potential $E^{\ominus}$ is the value of the electrode potential at 298 K when the metal is dipping into a solution of the metal cations at unit activity. The value of the electrode potential E at any other activity (or approximately, in dilute solution, at any other concentration) is given by the Nernst equation:

$$E = E^{\ominus} + (RT/nF) \ln [M^{n+}]$$

where R is the universal gas constant, T the temperature, F the Faraday and n the charge of the metal ion. At 298 K the equation approximates to

$$E = E^{\ominus} + (0.059/n) \lg [M^{n+}].$$

When an inert metal such as platinum is placed in a solution containing two different ions of the same metal (e.g. Fe^{2+} and Fe^{3+}), an electrode potential dependent upon the relative activities of the two ions is established, because the transfer of electrons between the inert metal and the two ions takes place initially at unequal rates. If n electrons are involved in the conversion of one metal ion into the other, the redox potential of the system is given by

$$E = E^{\ominus} + \frac{RT}{nF} \ln \frac{[\text{oxidized form}]}{[\text{reduced form}]}$$

the standard redox potential $E^{\ominus}$ being that obtained when both the reduced and the oxidized form of the metal ion are at unit activity. The more powerful the oxidizing ion, the more positive is the redox potential.

In using electrode potential data, it is important to remember that they are essentially equilibrium data and in themselves convey no *kinetic* information.

A more detailed treatment can be found in Chapter 4 of J. Robbins: *Ions in solution (2): an introduction to electrochemistry* (OCS 2).

Answers to problems

1.1. Iron(III) hydroxide is very much more insoluble than iron(II) hydroxide and this displaces the equilibrium reaction on to the iron(III) side.

1.2. The geometry of the ligand molecule will result in tetrahedral coordination.

1.3. 0·001206 mole of the complex produces acid that neutralizes 0·00345 mole of sodium hydroxide. Hence the complex is $[Cr(H_2O)_6]^{3+}(Cl^-)_3$.

1.4. The mercury(II) iodide dissolves with the formation of the tetraiodomercurate(II) ion, and the total number of particles in the solution is thereby decreased, resulting in an elevation of the freezing point:

$$HgI_2 + 2I^- \longrightarrow [HgI_4]^{2-}.$$

1.5. Both iron and copper have two readily accessible oxidation states, the higher of which is sufficiently strongly oxidizing to convert iodide ions to iodine, although in the case of iron the reaction is not very fast.

2.4. CFSE favours the square-planar arrangement, but the repulsions present cause the nickel complex ion to take up a tetrahedral configuration. The platinum atom is larger, and the repulsions are correspondingly less severe.

3.1. There are three points to make: the $E^\ominus$ values are measured at 25°C, and for molar solutions of the ions, and the escape of chlorine gas from the reaction mixture pulls the equilibrium almost completely to that side.

3.2. $+3, +2$.

3.3. The relevant redox potentials are 1·33 V for dichromate, 1·36 V for chlorine and 1·51 V for permanganate.

3.4. Distil the mixture with dichromate and concentrated sulphuric acid; collect the distillate in aqueous alkali and then test the resultant solution for chromate. Under these conditions, a positive result can arise only from the formation of chromyl chloride.

3.5. The values of E at the respective pH are 1·51, 1·32, 1·13, and 0·94. The $E^\ominus$ values for chlorine, bromine, and iodine are respectively 1·36, 1·07, and 0·54 volts.

3.7. Typical Jahn–Teller distortion for a d^4 system.

3.9. The oxidizing power of chromium(VI) is dependent upon the acidity of the solution. The presence of the acidic complex ion hexaquoiron(III) increases the oxidizing power when the two are mixed.

3.10. The pink colour is due to the octahedral aquo-complex $[Co(H_2O)_6]^{2+}$, whereas the blue is due to tetrahedral complexes such as $CoCl_4^{2-}$ which obviously depend upon the nature of the anion.

3.11. In the vapour phase there is practically complete dissociation of mercury(I) chloride into the mercury(II) chloride and mercury.

3.12. Answer:

$$E_{mercurous} = 0·789 + \frac{0·059}{2} \lg [Hg_2^{2+}] \text{ volts}$$

$$E_{mercuric} = 0·854 + \frac{0·059}{2} \lg [Hg^{2+}] \text{ volts}$$

At equilibrium in the presence of mercury metal,

$$E_{mercurous} = E_{mercuric}$$

Hence

$$\lg \frac{[Hg_2^{2+}]}{[Hg^{2+}]} = \frac{0{\cdot}065}{0{\cdot}059} \times 2$$

and so the equilibrium constant $\dfrac{[Hg_2^{2+}]}{[Hg^{2+}]}$ is about 160.

Reading list

J. KLEINBERG, W. ARGERSINGER, and E. GRISWOLD, *Inorganic chemistry*, Heath, Lexington, Mass. Good descriptive book, especially strong on electrode potentials. The theoretical side is unbalanced with too much emphasis on VB theory.

E. CARTMELL and G. W. A. FOWLES, *Valency and molecular structure*, 3rd ed., Butterworths, London (1966). This book gives a thorough treatment of atomic structure and the various applications of quantum theory to the problems of chemical bonding. The stereochemistry and bonding in the compounds of the elements are then discussed.

F. A. COTTON and G. WILKINSON, *Advanced inorganic chemistry*, 3rd ed., Interscience Publishers (1972). This is the best single textbook on inorganic chemistry, giving an excellent account of the principles of chemical combination with a more detailed treatment of the chemistry of individual elements.

C. S. G. PHILLIPS and R. J. P. WILLIAMS, *Inorganic chemistry*. Clarendon Press, Oxford (1966). This is an excellent advanced text on inorganic chemistry in two volumes. It is especially good in emphasizing periodic trends in chemical behaviour. The treatment is very rigorous but the book can be recommended for the above-average undergraduate.

D. P. GRADDON, *An introduction to coordination chemistry*, Pergamon Press, Oxford, 2nd edn. (1968). An interesting monograph with several unusual examples of coordination complexes. Useful background reading.

L. E. ORGEL, *An introduction to transition metal chemistry*, Methuen, London, 2nd edn. (1966). One of the earliest monographs on ligand field theory. Somewhat dated now, but a lucid, well-written account.

T. MOELLER, *Inorganic chemistry*, Wiley, New York (1952). The treatment of transition elements is rather brief, but there are several good chapters on complexes.

C. F. BELL and K. A. K. LOTT, *Modern approach to inorganic chemistry*, Butterworths, London, 3rd edn. (1972). The treatment of transition elements is again rather brief but there is a good variety of interesting complexes and unusual oxidation states discussed.

R. B. HESLOP and P. L. ROBINSON, *Inorganic chemistry*, Elsevier, Amsterdam, 3rd edn. (1967). Correctly described on the title page as "a guide to advanced study". It is inevitably a bit brief on detail.

A. L. COMPANION, *Chemical bonding*, McGraw-Hill (1965). An elementary treatment of bonding with some straightforward and useful exercises at the end of the chapters.

Related books in the Oxford Chemistry Series

J. ROBBINS (1972) *Ions in solution 2: an introduction to electrochemistry*. This constitutes a basic course in electrochemistry, and covers electrode potentials in some detail.

R. J. PUDDEPHATT (1972) *The periodic table of the elements*. An introductory survey of inorganic chemistry.

G. PASS (1973) *Ions in solution 3: inorganic properties*. This describes how ions dissolve and how they behave in solution, with emphasis on thermodynamics.

E. B. SMITH (1973) *Basic chemical thermodynamics*. A first course in thermodynamics from an approach of mechanical analogies.

C. A. COULSON (1973) *The shape and structure of molecules*. An introduction to the ideas of bonding theory.

Index

0336

SI units

Physical quantity	Old unit	Value in SI units
energy	calorie (thermochemical)	4·184 J (joule)
	*electronvolt—eV	$1·602 \times 10^{-19}$ J
	*electronvolt per molecule	96·48 kJ mol^{-1}
	erg	10^{-7} J
	*wave number—cm^{-1}	$1·986 \times 10^{-23}$ J
entropy (S)	eu = cal g^{-1} °C^{-1}	4184 J kg^{-1} K^{-1}
force	dyne	10^{-5} N (newton)
pressure (P)	atmosphere	$1·013 \times 10^{5}$ Pa (pascal), or N m^{-2}
	torr = mmHg	133·3 Pa
dipole moment (μ)	debye—D	$3·334 \times 10^{-30}$ C m
magnetic flux density (H)	*gauss—G	10^{-4} T (tesla)
frequency (v)	cycle per second	1 Hz (hertz)
relative permittivity (ε)	dielectric constant	1
temperature (T)	*°C and °K	1 K (kelvin); 0 °C = 273·2 K

(* indicates permitted non-SI unit)

Multiples of the base units are illustrated by length

fraction	10^9	10^6	10^3	1	(10^{-2})	10^{-3}	10^{-6}	10^{-9}	(10^{-10})	10^{-12}
prefix	giga-	mega-	kilo-	metre	(centi-)	milli-	micro-	nano-	(*angstrom)	pico-
unit	Gm	Mm	km	m	(cm)	mm	μm	nm	(*Å)	pm

The fundamental constants

Avogadro constant	L or N_A	$6·022 \times 10^{23}$ mol^{-1}
Bohr magneton	μ_B	$9·274 \times 10^{-24}$ J T^{-1}
Bohr radius	a_0	$5·292 \times 10^{-11}$ m
Boltzmann constant	k	$1·381 \times 10^{-23}$ J K^{-1}
charge of a proton	e	$1·602 \times 10^{-19}$ C
(charge of an electron = $-e$)		
Faraday constant	F	$9·649 \times 10^{4}$ C mol^{-1}
gas constant	R	$8·314$ J K^{-1} mol^{-1}
nuclear magneton	μ_N	$5·051 \times 10^{-27}$ J T^{-1}
permeability of a vacuum	μ_0	$4\pi \times 10^{-7}$ H m^{-1} or N A^{-2}
permittivity of a vacuum	ε_0	$8·854 \times 10^{-12}$ F m^{-1}
Planck constant	h	$6·626 \times 10^{-34}$ J s
(Planck constant)/2π	$\hbar$	$1·055 \times 10^{-34}$ J s
rest mass of electron	m_e	$9·110 \times 10^{-31}$ kg
rest mass of proton	m_p	$1·673 \times 10^{-27}$ kg
speed of light in a vacuum	c	$2·998 \times 10^{8}$ m s^{-1}

$\ln 10 = 2·303$ $\ln x = 2·303 \lg x$ $\lg e = 0·4343$ $\pi = 3·142$
$R \ln 10 = 19·14$ J K^{-1} mol^{-1} $RTF^{-1} \ln 10 = 59·16$ mV at 298·2 K